UK
Tel: +44 (0) 1372 802080
Fax: +44 (0) 1372 802079
E-mail: publications@pira.co.uk
http://www.pira.co.uk/

ISBN 1 85802 262 2

Typeset in the UK by Lewis Marshall, Epsom, Surrey
Printed and bound in the UK by T J International,
Padstow, Cornwall

Contents

List of tables

List of figures

Part 1 Factors influencing material selection

1
The selection and use of packaging materials

Packaging is big business. Throughout the world large quantities of material are used for the production of packaging. One estimate, from the World Packaging Organization, values material usage alone to be over 1350 million tons, with a estimated annual value of over US$475 billion.[1]

In addition to the materials used, large amounts of other resources are employed in the extraction, purification and processing of packaging materials. Significant quantities of energy, mostly in the form of fossil fuel, are used. Resources are required for the filling, closing and disposal of packages.

Packaging also facilitates the movement of other materials in business, commerce and trade. Every product, from food and consumer products to building materials and auto parts, is shipped or sold in a package. Many products require the use of a series of packages during their transformation from raw materials to finished goods.

A culture's packaging needs are related to its resources, demographics and technology.

The spread of packaging consumption shown in Table 1.1 reflects the degree of affluence and innovation in the more developed countries of the USA, Europe and Asia. These are the three main consuming regions in the world, although consumption in other countries is increasing with development.

Packaging is an international activity and the exchange of technical information has never been better. It does not take long for

Table 1.1 Major world packaging suppliers, 1994

Rank	Country	GDP US$ billion	Packaging production US$ million	Packaging production % GDP
1	USA	6638.2	95 862	1.4
2	Japan	4651.1	69 329	1.5
3	Germany	2041.5	24 851	1.2
4	France	1318.9	16 925	1.3
5	Italy	1020.2	15 518	1.5
6	UK	1013.6	13 185	1.3

Source: Howkins, M *World Packaging Statistics* Pira International (1997) p 4

any significant development in one part of the world to become known to interested practitioners in all others. In view of the vital role of packaging in improving the quality of life and reducing the losses of food and other products in developing countries, such international exploitation of improved technology is essential.

There are four basic packaging materials: glass; metal; plastics; and wood-based materials (including paper and paperboard). Within these four classifications there are many variations, each with a unique set of properties.

Paper and board is the largest single type of material used for packaging — 34% as shown in Table 1.2. Such estimates of the world's consumption of packaging materials are necessarily

Table 1.2 World packaging production by value and tonnage, 1995

	Estimated annual value of packaging		Estimated annual quantity of packaging	
	US$ billion	%	Million tonnes	%
Paper and board	160	34	500	37
Plastics	140	30	300	22
Metal	120	25	150	11
Glass	30	6	400	30
Others	25	5	–	–
Total	**475**	**100**	**1350**	**100**

Source: World Packaging Organization, cited by Howkins, M *World Packaging Statistics* 1997 Pira International (1997)

imprecise, and are likely to exclude many of the traditional materials which are still used to a significant extent in certain areas of the world, for example re-used paper and natural materials such as banana leaves.

Plastics are a close second, representing 30% (by value) of worldwide packaging production. In recent years plastics have been gaining market share at the expense of all other materials. Plastics are the youngest of the packaging technologies and are still very much on their upward growth curve.

The properties of plastics have progressed a very long way since the first one was developed over 100 years ago and most of this has occurred since the end of the Second World War. Plastics are now available which have the physical strength of steel, the temperature resistance of aluminium, the printability of paper, and barrier properties approaching glass. Although some of the more specialized grades developed for engineering applications are too expensive for packaging use, there is little doubt that some of the materials now being developed will allow for wider exploitation. This book explores the commodity packaging plastics as well as some specialized types which are finding new packaging applications.

The choice of packaging materials depends on the characteristics of the product and the expected packaging performance. The purpose of packaging is to protect, contain, identify and promote its contents — as well as delivering them in a useful form — often for a single use. Packages are also expected to facilitate product use including being easy to open, dispense, reclose and discard. Environmental concern over the extent of packaging consumption and disposal is also a major factor considered by packaging professionals seeking to conserve materials and reduce solid waste.

The purpose of this book is to provide a survey of packaging materials. It describes each packaging material, discusses the material's properties and applications, and indicates disposal options. It focuses on the efficient use of materials, because reducing and conserving packaging also make good economic sense.

Strategies to reduce and conserve

There are a number of approaches that may be taken to reduce the cost and amount of packaging material used, including substitution of materials, enhancing material performance, reducing scrap and increasing process efficiency.

Although some advertising-oriented strategies may increase materials usage in order to provide more product features or advertising space (and thereby increase sales), the aim of materials reduction is economy. Every penny saved in packaging costs multiplies into profit because of the high-volume production of most consumer goods.

The first approach is the rational evaluation of substitutable materials. The history of packaging is the story of a progression of materials, from the skins, baskets and pots of primitive hunter/gatherers to the colourful plastic forms that prevail in a modern supermarket.

Every new material is seen as a potential substitute for one or more of the existing types in use. Disposable packaging has always been made from the lowest-cost materials available to a culture. As new, lower cost materials are commercialized, packagers have found new applications for them.

Sometimes there is a perfect match and substitution rapidly occurs. For example (as described in Chapters 6 and 11), oriented polypropylene film rapidly replaced many cellophane applications because its properties were superior, it runs well on the same equipment and it is much less expensive.

In other situations there are limitations which make only partial substitution feasible. Plastics replacing glass and metal for processed foods is a good example, since food packages need better barriers and heat resistance than most plastics alone can provide. When such a competitive threat arises, it provides a spur to development of both the new and the old materials.

Of the four main groups of materials, paper-based and plastic have experienced the most steady growth over recent years. Metal and glass are losing share, having been replaced by plastics or paper/plastic combinations for many applications. One purpose of this book is to highlight the substitutability of materials used for packaging.

A second approach to materials development is to enhance the performance of existing materials, making it possible to use less material. For example, plastics manufacturers are continually researching better formulations of commodity plastics, like the use of new metallocene catalysts in the manufacture of polyethylene (described in Chapter 6), improving almost all of its properties.

Another way to enhance existing materials is by combining the best performance aspects of a number of different materials, each contributing its special set of properties to the final result. Examples include plastic coating to strengthen and protect glass, and paper/foil/plastic laminations to lengthen the shelf-life of food products. Chapter 12 outlines some methods for enhancing packaging materials' strength and barrier properties, two of packaging's most demanding attributes. Chapter 13 discusses material modifications such as coatings and material combinations.

A third economical approach is to improve materials utilization, reducing losses due to faulty production and making greater use of in-plant scrap. The relative importance of materials conservation varies among the different material types.

Table 1.3 shows the relative significance of raw materials costs. Metals are at the top of the scale — the materials account for between 75% and 80% of the final cost of a tinplate or aluminium container. Paper and plastic materials account for about 50% of their total container costs, and glass with its cheap raw materials (mostly sand) is at the bottom with 20–25%.

Table 1.3 Raw material cost as a percentage of container cost (in general)

Material	%
Metals	75–80%
Paper	50%
Plastics	50%
Glass	20–25%

Scrap utilization figures vary with the manufacturing process used and the level of technology involved. With metals and paper conversion, in-plant scrap has to be collected and returned to the primary material producer. Plastics and glass manufacturers, on the other hand, can usually put their scrap straight back into the process to produce new materials or containers. Likewise, all four materials can be easily recycled after use.

A fourth impetus for materials development is to keep up with demands for higher filling machine speeds at product manufacturers' plants. At the same time, manufacturers are developing mass customization strategies that favour short production runs with quick changeovers.

Sometimes an apparently minor aspect of a package can make a great deal of difference in its ability to run on machines. The development of cold seal adhesive is one of the best examples, because it allows horizontal form–fill–seal machines to be run at very high speeds and is now used extensively to pack confectionery products (as described in Chapter 14).

External factors influencing the choice of materials

There are a number of natural and social factors that influence a culture's choice — its supply and demand — of materials. Availability of natural resources and the state of packaging technology affect the supply of materials. Social and cultural norms, such as lifestyles and environmentalism, as well as trends in marketing and distribution affect packaging material demand.

Natural resources

Historically, people have used whatever materials were readily available and those which we have had the knowledge and technology to adapt. Thus packaging began with natural materials — gourds, animal skins and large leaves — progressed to easily worked materials like wood and clay, and then on to paper, metal and glass, and finally to plastics. Plastics are quite different from earlier materials in that they are not a simple conversion of an existing material, but involve the modification of the basic structures of chemicals to produce entirely new compounds that do not exist in nature.

The pattern of development, however, has been the same for each material, from the small-scale discovery through experiment to advanced-scale manufacturing, with the cost falling at each stage. For example, in 100 years (from the mid-1800s to the mid-1900s), canned food went from being a luxury good to the common and inexpensive product that it is today. In that time, the processes for making tinplate, seaming cans and filling them became vastly more efficient.

Most of the material resources used for packaging are either renewable (wood and vegetable fibre for boxes and paper-based packaging) or very plentiful (sand for glass, clay for ceramics, iron ores for metal, and, much later, bauxite for aluminium). Even the finite resource of petroleum, used as feedstock for plastics, is relatively plentiful considering the small amount of the world's oil production that is used — very efficiently — for making plastics.

The other primary resource used for the manufacture of all forms of packaging is energy. Four main sources are used: fossil fuel, short-term renewable forms of energy such as wood burning, 'free' energy (solar, tidal, wind and hydro-electric) and nuclear energy. Different materials require differing energy inputs.

Suffice it to say that all non-renewable resources based on minerals present in the earth can be recovered, however much they have been transformed during their use, provided that sufficient energy is available. This has led some commentators to suggest that in the long term the true cost of all forms of packaging will be determined by the amount of energy employed in their manufacture and use.

Packaging also affects disposal resources. Packaging is about 30% of the weight in the domestic dustbin or bag. Residential community opposition to landfills and incineration has made such options more costly. Recycling systems, while reducing the amount of waste, have sometimes produced worse environmental effects than disposal, including air and water pollution and high energy use.

In the industrialized world, criticism has been aimed at the packaging industry by some ecologists, environmental pressure groups and consumers. The most frequent charges are that packaging makes excessive and wasteful use of resources, adds to the burden of waste disposal, and uses large amounts of energy including fossil fuel as feedstock for plastics.

In recent years, there has been a growing call for all products, including packaging, to be seen to be 'green' (i.e., sensitive to the ecological needs of the planet). The packaging industry has not always responded to the criticism in a productive manner. The manufacturers of one material (like plastic or paper) have sometimes blamed the manufacturers of competing materials. One effect of such internecine warfare has been to harden the attitude of the activist protection groups against the whole industry. In some countries, notably in Europe and Japan, recycling has now become mandated at any cost.

A more productive approach is to understand the criticisms and to reduce the burden of packaging whenever practical. 'Over' packaging is costly, and is rarely a successful strategy anyway. Producers are well advised to reduce the resources used for packaging, and packaging reduction has long been a successful strategy for cost reduction.

All packaging materials can, technically, be recycled. Table 1.4 shows the relative recycling rates of municipal solid waste (MSW) in the USA (where recycling is based on economics, rather than legal mandate). Corrugated fibreboard has, by far, the best recycling rate in this table, of over 30%.

Table 1.4 Packaging material recycling rates in the USA

Material			Recovered	
	Tons million	Percentage of MSW	Tons million	Percentage of total
Glass	12.1	5.8	3.1	6.3
Steel	3.1	1.5	1.6	3.2
Aluminium	2.1	1.0	1.2	2.3
Corrugated	28.4	13.6	15.7	31.9
Paper and paperboard	9.4	4.5	1.4	2.8
Plastics	9.5	4.5	0.7	1.4
Wood	10.2	4.9	1.4	2.9
Miscellaneous	0.2	0.1	–	–
Total	**75.0**	**35.9**	**25.1**	**50.8**

Source: Environmental Protection Agency *Characterization of Municipal Solid Waste in the United States: 1995 Update*

The economics of recycling vary by material. The difficulties and costs are associated with the need to sort materials so that they can be homogeneously recycled. Corrugated fibreboard is economical to collect in large homogeneous amounts from retailers and warehouses, and is economical to reprocess. However, for some other materials, like multi-material barrier film mixed in minute amounts with other household packaging, the collection, sorting and reprocessing often uses more resources than it saves.

Furthermore, there is a conflict between the easy recycling of single-material packages and economic use of multi-material systems. Multi-material packages, although providing superior protection at low cost, can seriously impair the economics of scrap re-use and recycling.

The demand for more package recycling — legislated in some parts of Europe and Japan — can have an important effect on the selection of packaging materials. Materials manufacturers and designers can greatly affect the ease with which packaging can subsequently be collected, identified and recycled to produce new items, but it may be at the cost of using materials less efficiently.

There have been some who have promoted biodegradable and photodegradable plastics for the amelioration of the problem of litter and for packages that are ultimately landfilled, even though little biodegradation actually occurs in modern sanitary landfills. Biodegradable packages have a number of problems, including the fact that the degradation may occur too soon, during the useful life of the package.

In the past, cost has been the primary method for assessing packaging materials. Environmentalist critics have argued that traditional internal cost analysis does not include the external costs of disposal and pollution that are borne by society.

Life-cycle analysis (LCA) is a new technique that is gaining power as a method to evaluate the burden of packaging on the environment. It takes into account the resources required for raw material extraction, packaging material conversion, package filling, product distribution and packaging disposal, recycling or re-use. It considers the renewability of natural resources as well as pollution effects. This 'cradle to grave' approach can be useful for comparing alternative packages.

Demographic and market shifts

The market shift that has most affected packaging is the change in the twentieth century from small retail stores with helpful service assistants to self-service shopping in large hypermarkets. In most retail settings today, the package is expected to be a 'silent salesman', attracting attention, linking retail shelves to media advertising, and conveying important information such as nutritional content and warnings.

Packaging facilitates self-service shopping and the use of all kinds of consumer products, from food to shampoo to toys. Attractive materials, like holograms, are used to lure shoppers, and easy-to-use features like easy openability and reclosability tend to make shoppers brand loyal.

But the advertising function of packaging has been blamed for excessive use of packaging materials, as some manufacturers have exploited shelf-facings by making packages as large as possible. Increasingly, consumers are not fooled by this strategy. They complain that packaging adds unnecessarily to product costs, that it can limit consumer choice, that it may be used in a deceptive manner, and that it appears as a major constituent of litter and landfills.

Packaging decision-makers must weigh the benefits of the in-store billboard effect, which in the past has been shown to attract customers, against the growing demand for more efficient use of materials. In the past few years, product concentration and packaging reduction strategies have been gaining market share.

The second market change with great significance for packaging is demographic. There have been huge changes in social patterns in recent years, and while these are at different stages in various parts of the world, the general trend is similar, albeit at different paces.

The largest segment of the packaging industry deals with foods. Food manufacturers are the most sensitive to changes in lifestyle and demography, since the changes affect, and are affected by, the way that we eat and prepare food.

More women now are employed outside the home, and so there is an increasing demand for greater convenience in food preparation. More of the population consists of single-person households, and family meals have become less common, increasing the demand for easy-to-prepare single meals. Coupled with the dynamic growth of the microwave oven, these changes have led to huge growth in demand for part- or fully prepared meals and snacks.

Such convenient refrigerated meals necessitate different forms of packaging, often involving very demanding performance to increase shelf-life. Longer shelf-life can be achieved by using materials with superior barrier properties (to water, oxygen, bacteria). In the case of fresh produce, materials are selected to modify the atmosphere inside the package and slow the product's respiration rate.

The demand for microwavable packaging has also stimulated the development of heat resistant materials that are dual-ovenable, that is, can be heated in either a microwave or a conventional oven. Since microwave ovens cannot brown food, metallized 'susceptor' materials, that create a local microwave energy hot spot, have also been developed.

Another important demographic trend is the aging population in many parts of the world. Older consumers have special needs for product information and ergonomic designs, especially with regard to medical products which are more often used by older people.

The near future holds a third marketing trend that has significance for packaging. Consumers' desire for convenience is increasingly being met by home deliveries of products ordered by catalogue or computer. Clothing and home furnishings are already widely ordered by mail in some countries (especially the USA). Home delivery of fast-moving consumer products is being tested in many places, and is predicted to grow.

Home-delivered products will stimulate some interesting packaging changes. The package will play a different role in stimulating shopping choices. Packages function less to attract shoppers for the initial sale, but will be called on more to reinforce purchase decisions and encourage the shopper to buy again. For home delivery of foods, shipping containers may need to be insulated or refrigerated.

Home deliveries and other logistics advancements stimulate other changes in shipping container design. The new supply chain management approach to logistics requires that shipping containers facilitate product flow through sorting operations. With shorter supply chains, there is less warehousing, including storage and order picking. Orders are shipped directly from the manufacturer to the retailer, with simple sorting operations between. Such just-in-time systems can create opportunities for more efficient shipping containers. For example, they can often justify in economic terms the use of returnable containers.

For direct delivery, whether to a home, business or retail store, shipping containers need to be easy to sort using automation. In such cases, packages need to have standard dimensions and materials (generally corrugated fibreboard boxes or reusable plastic totes), and bar codes need to be in a standard position.

Global technological developments

Packaging is a truly international activity. Marketing information technology and logistics infrastructures now allow sourcing to take place on a global basis, with packaging being done at the most rational point in the supply chain.

The packaging industry, within countries and across borders, is experiencing a long period of restructuring and consolidation, leading to polarization of both markets and companies. International packaging supply companies have developed, leading to greater standardization of materials across countries. At the same time, there are opportunities for small suppliers to exploit fast-growing niche markets with high added value potential.

Although social, climatic and market situations differ widely, technology is readily transferable around the world. Those involved in pure packaging innovations consider the world as a source of inspiration and information. Many would agree that Japan represents the major centre of sheer innovation, while the USA and Europe concentrate on fewer but larger-scale developments.

In Japan, packaging is both an art form and a science. Owing to their cultural background, Japanese consumers seem especially to appreciate high quality and novelty packages. Their gift-giving tradition emphasizes high quality presentation, and packaging is as important as the item itself. In order to achieve pure novelty, often for its own sake, manufacturers will offer a bewildering range of packs to tempt the purchaser into buying, especially where market pressures are high, as in the beer and soft drinks sectors. While other markets may not require this level of sophistication — or may consider it to be positively wasteful — the technical developments such as the self-heating sake can have found uses in other packaging situations like the self-heating dinner for drivers.

Packaging plays a significant role in a nation's economic development. It increases the protection of domestic goods and enables their export into new markets. Generally, the expenditure on packaging is much lower in less developed countries. Most materials and graphics are less sophisticated, but the packaging that exists is generally economical, material recovery systems are efficient, and there are many creative packaging applications using indigenous materials and appropriate technology. When a country decides to export its goods, it usually finds the need for a package redesign to increase protection and improve graphic communication.

Requirements for selection of packaging materials

Given the wide range of substitutable packaging materials, choices must be based on the economics and performance required. Such requirements must take into account the nature of the product, expected packaging performance, decoration and advertising function, method of manufacture, legal and safety needs, and, most importantly, total cost.

Material performance requirements

Materials and pack styles should always be selected on the basis of their performance. In considering the various options, some will clearly be unsuitable while others will have to be evaluated in terms of the degree of suitability.

The first factor to consider is the product and its characteristics to determine the kinds of protection required. It is also important to consider the nature of the marketing channel to determine the time and other constraints involved.

In general, food has the most specific protective packaging requirements. Dry food products require protection from moisture which can make them soggy. Food containing fat requires protection from oxygen which causes rancidity. Fresh produce requires protection from respiration which causes rot. All food needs to be protected from contamination, bacteria and vermin which can cause illness and disease.

Clearly the barrier properties of food packaging materials are an important consideration. The longer (in time) the marketing channel, the more important the barrier becomes. This book explores the barrier properties of packaging materials. Chapters 6–11 examine the barrier performance of the various polymers, and Chapter 12 provides a comparison among plastics. Chapters 3 and 4 show why the barrier performance of metal and glass has made these two materials the most popular food packaging for over 100 years.

Food and other kinds of products also need protection from mechanical damage. There are several sources of mechanical damage in distribution. Impacts occur during handling, vibration occurs during transit and compression results from stacking. It is important to understand the specific distribution system and the hazards involved.

It is also important to understand the specific product's response to impacts, vibration and compression. Potato crisps need protection from breakage. Building materials and ready-to-assemble furniture require edge protection in order to be able to connect smoothly. Electronic equipment requires protection from impacts, vibration and electrostatic discharge. The properties of cushioning and other shipping container materials are described in this book. Chapter 12 compares the mechanical properties of plastic materials.

Furthermore, the package should not interact with its contents. This is most important for food products where lead (from soldered seam cans) and migrating constituents of plastics have been found to be contaminants and a risk to health. The package should not impart colour or cause abrasion. In the case of chemicals, the package should not dissolve, stress crack or break down during contact.

Once the protection requirements are defined, it is a reasonably straightforward matter to compile a list of feasible materials. The next consideration is how the package will be filled and sealed. Generally, the final choice depends on the process being compatible with some existing equipment and processes. Some factors include:

- the ability to be heat sealed or closed some other special way;
- the capability to withstand high or low processing, filling or storage temperatures;
- compatibility with existing packaging equipment; and
- safety and compliance with legal (e.g., hazardous material) requirements.

Each material chapter, especially Chapters 6–13 which discuss plastics and laminates, includes heat sealing and heat processing considerations, as well as issues of chemical compatibility.

Pack styles

Materials are converted into packages in various ways. Some, like cans, bottles and boxes, are supplied in a ready-made form.

Others are produced in-line on the filling machine, such as form–seal–fill types made from flexible materials and cartons made from cartonboard reels or pre-cut blanks. The selection of materials and form of packaging for any specific products are closely related. Each material can be fabricated into a number of different forms. Some of the options are shown in Table 1.5. Plastics are the most versatile group of materials, and can be made into every single form.

Table 1.5 Packaging materials and their forms

Material	Possible forms											
	1	2	3	4	5	6	7	8	9	10	11	12
Aluminium	x	x	x	x	x	x			x			x
Glass	x			x	x							
Tinplate/tin-free steel	x	x	x	x	x							
Cartonboard/corrugated		x	x	x		x		x			x	x
Paper		x	x	x		x		x		x	x	
Plastics	x	x	x	x	x	x	x	x	x	x	x	x
Flexible laminate		x	x	x		x	x	x	x	x		x

Forms

1 Bottles and jars (not precisely defined, the difference is mainly one of geometry and aperture size)
2 Open-top cylinder cans and drums
3 Square or parallelepipedal (brick-shaped) packs
4 Irregularly shaped three-dimensional packs
5 Aerosol and other pressure dispense packs
6 Bags, sachets and sacks
7 Chub (large sausage-shaped containers)
8 Blister, skin and similar carded packs
9 Collapsible tubes
10 Heavy-duty sacks
11 Transit cases and trays
12 Intermediate bulk containers

Table 1.5 is based on packages made from a single material. It is increasingly common to combine materials to offer the best combination of their individual properties. For example, a paper/aluminium/polyethylene flexible laminate uses the aluminium component for an absolute barrier to gas and water vapour; polyethylene provides a heat-sealing medium as well as increasing the mechanical strength and filling tiny pinholes; and paper provides a low-cost form of mechanical stiffness and an excellent printing surface. Such materials, combined in various proportions, provide some of the best known flexible materials, ranging from drink boxes to retort pouches, as described in Chapter 13.

Methods of manufacture

The range of package manufacturing options is wider than it may at first appear. The most commonly used methods of manufacturing for each of the materials are as follows:

- *Aluminium metal:* Impact extrusion, deep drawing, cold forming, spinning and, when rolled into thin foils, any flexible or carton manufacturing process.

- *Tinplate or tin-free steel:* Rolling, bending, seaming and multistage stamping.

- *Glass:* Moulding direct from furnace into final shape, ribbon blowing.

- *Paper and board:* Moulding from pulp, corrugating, hot pressing, spiral winding, cutting, creasing, adhering, and erecting by mechanical interlocking. When paper components are combined with plastics, their versatility is vastly increased.

- *Plastics:* Extrusion or injection blow moulding, thermoforming, sheet fabrication and adhesion or heat/radio frequency sealing, injection moulding, film production, laminating, fibrillation, weaving, perforating and so on.

The properties of materials can be much influenced by their processing. Today, with the wide range of basic materials available for packaging, which can be used either alone or in combination, modifying their nature by some processing technique may sometimes provide even broader spectra of properties. The treatment may be carried out at the time of manufacture or as a post-fabrication operation. Examples include orientation (stretching), surface treatment, foaming, cross-linking by irradiation, crystallization, cold rolling and annealing, as well as blending and combining by various means.

Decoration

Another element in the selection of materials is the form of decoration to be used. All too often this is left until the decision on the material type has been made, and then any inherent restrictions arising from that choice have to be dealt with. Table 1.6 shows the decorating options suitable for various packaging materials. All forms of packaging can be labelled, but paper and plastic are the easiest to print directly.

Table 1.6 Decorating options for various packaging materials

Method of achieving decorative effect	Suitable for containers made from
Self-colour — integral in base material	Glass, plastics, paper
Embossed or moulded design	Plastics, metal, paper
Labels (paper or plastics) adhered or shrunk on	All forms of packaging materials
Labels adhered in forming mould	Plastics only

The two most common package printing processes are flexography and gravure. Flexography uses an elastomeric raised relief printing plate and fast-drying inks. It can be used to print on a wide range of substrates, including plastic films, foils, coated and uncoated paper, paperboard, and corrugated board. Gravure uses an engraved cylinder, and provides more consistent results. It is used on smooth substrates and for longer production runs, especially for paperboard cartons.

Packages can also be decorated using letterpress, lithography, silk screen and hot stamping processes. Letterpress printing uses a rigid relief image plate of metal or photopolymer material, requiring a very smooth substrate such as coated paper. It is used for labels, blister cards and decals. Offset letterpress involves first printing onto a blanket that, in turn, transfers the image to the substrate. It is used mostly for the printing of two-piece cans. Offset lithography separates the image and non-image areas of the print roller chemically, rather than by relief, using paste ink that adheres only to the image area. The image is then transferred to an offset rubber blanket which can print on fairly rough surfaces, but prints poorly on plastics.

Silk screen printing is used for small quantities of rigid parts, like cosmetics bottles, where a luxury image is desired. Hot stamping is used for metallic and other effects. Ink jet printing uses a computer to generate ink droplets in a pattern, and is increasingly used for printing text on demand for date/lot codes and shipping container identification.

New developments are occurring at a particularly fast rate in the printing field. Imaging technologies, developed for the graphics and computer industries, are certain to find applications in 'on demand' decoration of packaging, enabling shorter production runs and a quick means to product differentiation.

Economics

A package design should make optimum use of materials, be easy to use, attract and promote sales, provide information to the purchaser, and be easy to discard after use. All of these criteria should be met in the most economical way. Establishing agreed criteria for these different aspects is not easy, and measuring them is even more difficult.

It is important to note that the economics of different manufacturing, production or decorating methods should always be compared on the same basis, and should include benefits as well as costs. It cannot be stressed too strongly that when considering the economics of a packaging material, pack or system, it is the total cost which must be considered.

All too often, materials or packs are chosen on the basis of their purchase price only. It is often forgotten that associated costs, such as filling, storage, transit, damage, and retail efficiency, can all have a major impact on the overall economics. For example, a glass shampoo bottle may well be less expensive than the plastic alternative from a purchasing viewpoint, but the fact that the glass is heavier, more bulky and more vulnerable than plastic can result in higher transport costs, greater damage costs and a higher incidence of injuries (and hence higher liability costs).

Furthermore, packaging affects the cost and demand for every product and factor of production, from food to building materials to auto parts. All products are packaged in some way and may have been repacked several times before use. Even at 5–10% of the retail price, packaging costs can amount to hundreds of millions of dollars for a multi-billion dollar food or beverage firm.

Nevertheless, it is also important to recognize that packaging is more than a cost centre. Packaging directly affects sales and profitability. A packaging innovation may have a higher purchase price than the container it replaces but can increase profits by adding value, increasing product quality, reducing damage and improving efficiency of production and distribution.

This book, however, is about packaging materials and only incidentally deals with such added complexities. The chapters explore each packaging material in turn, summarizing their properties and applications. Taking a somewhat historical approach, Part 2 has three chapters about traditional packaging materials: wood, paper, paperboard, corrugated fibreboard, metal and glass. Part 3 explores synthetic (plastic) materials in eight chapters and Part 4 discusses composite materials, emphasizing flexible packaging, and ancillary materials ranging from adhesives to cushioning. Endnotes are provided for specific references to new technology which may have been described only in recent journal articles. The book ends with a comprehensive bibliography of references and an index.

Endnotes for Part 1

1 Townshend, GK 'The packaging industry today' *Proc. International Packaging Conf.* Beijing, China, Packaging Technology Association (1996) pp 4–9, *cited in* Howkins, M *World Packaging Statistics 1997* Pira International (1997) p 1. It should be noted that such estimates are a matter of debate.

Part 2 Survey of traditional materials

2
Wood and paper-based materials

With the exception of clay and natural materials such as reeds and leaves, wood is the oldest material used for packaging. Examples of 5000 year old chests have been found in tombs of the old kingdom in Egypt. Later and better preserved specimens were found in Tutankhamen's tomb.

Paper has also been around for a long time. A crude form, papyrus, was originally produced in Egypt from a sandwich of fibrous plant materials. The modern papermaking process was developed in China over 2000 years ago. However, it was not until the middle of the nineteenth century that wood pulp began to be used for making paper on a large scale, and not until the twentieth century that paperboard and corrugated fibreboard became popular packaging materials.

Paper

Paper represents the largest proportion of materials used for packaging; 34% by value and 37% by weight (see Table 1.2 above). Most papers are used for flexible packaging applications, and much of it is coated or laminated with plastic.

Paper offers a number of benefits. It performs well at a low cost. It is stiff, opaque, easy to print and versatile. It has proven success in wrapping and labelling applications. In recent years its environmental profile has helped the material to resist continuing substitution by plastics to a greater extent than have metal and glass.

Paper can be made from many naturally occurring forms of cellulose (wood, cotton or grasses) which are macerated in water and laid out to dry under heat and pressure as a flat sheet. The water softens the outer surface of the cellulose fibres, which then fuse together when they make contact with other fibres using pressure or suction (or both) in a modern papermaking machine. The bond is made stronger and hardened during the calendering stage when the paper is driven through a set of heated metal rollers.

Today's packaging papers are made from softwood trees, especially spruce and pine. Other tree-derived sources include birch and eucalyptus, but seed hairs like cotton, fibres such as flax, jute and hemp, and grasses including straw, bagasse (sugar cane byproduct) and bamboo, as well as leaf structures such as esparto and sisal, all have the same cellulosic structure with its ability to form strong fibre-to-fibre bonds, making them suitable for paper. The composition of wood pulp, after stripping off the outside bark, is about 40–45% cellulose and a further 5–10% hemicellulose, plus lignin.

The world packaging paper and board industry is concentrated in Europe, Asia and North America (see Table 2.1).

Table 2.1 World packaging paper and board production, 1993–94

	Production		Imports		Exports		Apparent consumption	
	1993	**1994**	**1993**	**1994**	**1993**	**1994**	**1993**	**1994**
Europe	8 366	8 758	4479	4763	3530	4009	9 315	9 512
North America	3 483	3 567	406	501	486	557	3 403	3 511
Asia	3 791	4 529	985	1136	194	250	4 582	5 415
Australasia	31	25	4	5	6	6	29	24
Latin America	1 015	536	74	88	21	23	1 068	601
Africa	381	189	140	144	5	5	516	328
Total	**17 067**	**17 604**	**6088**	**6637**	**4242**	**4850**	**18 913**	**19 391**

Source: Howkins, M *World Packaging Statistics 1997* Pira International (1997) p 5

Papermaking process

There are a number of physical and chemical processing routes by which the raw material is converted into paper, all of which begin with a physical size reduction. This beating reduces the wood to fibres of desired length and softens their surfaces. If little or no further treatment is undertaken, a high yield, but low quality paper — mechanical grade — is the end product. Newspaper is the main use for this, and the characteristic browning and embrittlement which take place after it is exposed to strong sunlight is due to degradation of the lignin. Mechanical paper of this type is therefore suitable mostly for very short-lived products.

Most packaging papers are subjected to a chemical process after an initial mechanical breaking stage. This is a more gentle way to tease apart the fibres and activate their surfaces to initiate bonds. The ground-up wood is cooked in a solution of either acidic sulphite (to produce bleached kraft paper) or alkaline sulphate, to produce the stronger unbleached kraft paper. Since the lignin and much of the hemicellulose dissolves into the chemical solutions, both of these processes reduce yields.

Improved techniques in the manufacture of paper, especially the use of a hybrid process to produce chemo-mechanical pulp have raised the yield achievable, increasing the productivity of individual mills, reducing costs, and improving the environmental profile of the industry as a whole.

The properties of the finished product depend on the degree of beating and chemical digestion, plus the addition of other chemicals such as bleaching and sizing agents. Materials as different in texture and appearance as blotting paper, tissue and transparent glassine can be the end product, depending upon the process used.

At the end of the mechanical/chemical processes, a thin slurry of cellulose fibres (only about 1% by weight) in water remains. This is laid down on a moving fine mesh wire belt in the Fourdrinier process which is used to make most paper, or onto a circular mesh drum in the cylinder process which is generally used for paperboard since successive layers can be built up into a thicker sheet. The water drains through the mesh and a vacuum is then applied to compact the fibres and extract more water before the paper moves to the drying stages.

Paper thickness is dictated by the number of layers of fibre which are laid down via a series of head boxes through which the fibrous slurry is dispensed. Some paper manufacturers make use of this multiple-layer capability to produce grades having different colours or paper textures on the two faces for special effects. Similarly, it can be used to provide a paper made mainly from recycled fibres with a thin outer coating of virgin fibres to improve appearance. Another application is the production of the so-called oyster finish in which an incomplete skin of bleached white fibres forms an overlay on unbleached kraft to provide a mottled effect as the brown colour shows through the thinner areas of white.

The most extensive use of this multi-layer production ability is in the manufacture of cartonboard, which is really only thick paper. The definition of what constitutes paper and what is board is not universally agreed; generally structures less than 0.012 in. thick (12 'points' or 300 μm) are considered paper; the ISO standard marks the transition at 250 g/m^2.

Bleaching is an area of concern for its environmental effects. The use of chlorine-containing chemicals leads to the production of minute traces of dioxin in the water effluent, and even smaller amounts can remain in the paper pulp. Dioxin is harmful when ingested or in contact with skin. Although there has not been one single recorded incident of any harmful effects on humans from either source, there has been widespread concern. Scientists in the paper industry have refined detection methods to be able to identify amounts as low as parts per quadrillion, which are infinitesimally low. However, once detected at any level at all, concern can be aroused and paper mills worldwide have launched a comprehensive programme to reduce the use of chlorine in their processes.

There have been two responses to the bleach concern. The first is the greater use of unbleached paper, which also produces a stronger material. The second is to use alternative chemicals where bleaching is essential. Oxygen and chlorine dioxide are two of the main chemical routes being adopted; the latter significantly reduces the amount of chlorine emissions and the former does so completely.

There are several dry-end methods to modify the properties of paper by coating it with lacquer, aqueous dispersions, molten plastic or wax. Paper can easily be coated with any thermoplastic material by extruding a molten layer over its surface. Low density polyethylene is most frequently used, especially for heavy-duty barrier materials where it provides both a moisture barrier and heat sealability. Other plastics used include PVdC, nitrocellulose, acrylics and PET. This wide choice makes possible an equally wide range of materials having different performance for many applications. Such coatings improve the wet strength or barrier properties (or both) while retaining good printability and other benefits of paper.

Paper can also be laminated with other materials including aluminium foil, fabrics and non-woven fibres, and loose filament nets to improve mechanical and barrier properties.

Packaging grades

The furnish, proportion of fillers used, the degree of beating, the amount of filler, binders, sizes and so on, together with the operating variables of the particular papermaking machine and finishing processes, can be varied to produce many types of paper. The principal packaging papers, as well as their origin and uses are outlined in Table 2.2.

Table 2.2 Main packaging papers

Basic material	How made?	Weight range		Tensile strength	Properties and uses
		lb/ 1000 ft²	kg/ 1000 m²		
Kraft papers	From sulphate pulp on softwoods (e.g., spruce)	14–60	70–300	High	Heavy duty paper, bleached, natural or coloured; may be wet-strengthened or made water repellent; used for bags, multi-wall sacks and liners for corrugated board; bleached varieties for food packaging where strength required
Sulphite papers	Usually bleached and generally made from mixture of softwood and hardwood	7–60	35–300	Varies	Clean bright paper of excellent printing nature used for smaller bags, pouches, envelopes, waxes papers, labels and for foil laminating, etc.
Greaseproof papers	From heavily beaten pulp	14–30	70–150	Medium	Grease-resistant for baked goods, industrial parts protected by greases, and fatty foods
Glassine	Similar to greaseproof but super-calendered	8–30	40–150	Medium	Oil and grease-resistant, odour barrier for lining bags, boxes, etc., for soaps, bandages and greasy foods
Vegetable parchment	Treatment of unsized paper with concentrated sulphuric acid	12–75	60–370	High	Non-toxic, high wet strength, grease- and oil-resistant for wet and greasy food, e.g., butter, fats, fish, meat, etc.
Tissue	Lightweight paper from most pulps	4–10	20–50	Low	Lightweight, soft wrapping for silverware, jewellery, flowers, hosiery, etc.

Source: Paine, F *The Package User's Handbook* Blackie (1991) p 46

Brown kraft paper originated in Scandinavia. The word means literally 'strong' and it is used in multi-wall bags and other applications where strength is needed. Kraft paper is used to make corrugated fibreboard facings.

Clean, bright solid bleached sulphite (SBS) paper is white throughout and is generally highly sized for water resistance; it has excellent printing characteristics. It is used for small bags, pouches, labels and for foil laminating. Kraft and SBS can also be clay coated to improve the printing surface.

Greaseproof, glassine and parchment papers are made to resist oils. They are used as the lining layer in packages for dry and oily food products such as baked goods, dog food and butter. Lightweight tissue paper is used for wrapping.

Unconverted paper can be used as a flat material for wrapping, interleaving, stiffening and space filling. When converted into sachets, envelopes and bags, paper packages can be easily filled by hand or by machine. There is a wide range of adhesives available. Multi-wall paper bags are an ideal application for paper's unique properties since they combine the toughness of multiple plies (which may include a plastic ply for its barrier properties) with the excellent printability of paper.

In labels, laminated and coated structures, paper has proven its value by its ability to combine with all other materials to make optimum use of the properties of each. Paper adds the ability to use high quality printing on such multi-layer structures.

New developments

The success of paper as a packaging material has inspired scientists to invent variations on the theme. New processes to reduce energy consumption and new feedstocks ranging from plastics to waste products have been developed.

Papermaking has remained a water-based process since it was first invented. This fact makes energy a very important cost element in the economics of paper manufacture, since the water has to be progressively extracted from the initial slurry, at about 1% solids, to a dry paper with only 4–6% moisture content. The process is not particularly energy efficient, and for many years research has been carried out to devise a way to produce paper by a dry method, without the need to use suction and evaporation. Although the industry does add some resin sizing to the pulp to promote fibre adhesion, it has not been possible to make this the only bonding mechanism. Attempts to increase the surface area for fibre bonding have included the use of explosive steam under pressure to promote the separation of fibres.

There are a number of paper products that incorporate the use of plastic or resin. Organic resin binders like those used to make particle boards are expensive and such products have not been able to match the properties of natural paper. Blending polyolefin fibres into the pulp can provide a strengthening matrix of long fibres in the paper, but such papers require special treatment for creating the bond with the cellulose fibres since the polyolefins cannot be bonded by water.

Several all-synthetic papers have been commercialized for uses where strength, water resistance or 'printability' is required, and an extruded film or wood-based paper would not suffice. Tyvek (a DuPont tradename) is a very strong material made from heat-fused, long, high density polyethylene (HDPE) filaments, used for medical wrappers and security envelopes. Other attempts by companies like Oji Yuka have involved loading polyolefin and polystyrene films with inorganic pigments such as chalk, talc and titanium dioxide in order to achieve high quality printability. Mobil Plastics' Oppalyte (made from oriented polypropylene) combines many of the physical properties of paper but has the benefits of water resistance, heat-sealability and extremely low density. Metallized paper is becoming more widely used in cigarette packaging and for labels where curling problems have been experienced with foil/paper laminations.

Recycling is an important issue for the paper industry which has always used some recycled inputs. The most economical sources are used newspapers, magazines and corrugated fibreboard boxes, since these are easy to collect in large homogeneous amounts. Other papers used in packaging are more difficult to sort and collect, and many are contaminated, which can increase the cost of recycling beyond the level of economic feasibility.

Packaging paper can include recycled material in most applications. The use of recycled fibres is growing, owing to an increase in paper recycling worldwide. It is important to note, however, that recycled fibres are shorter than virgin paper fibres, and that paper made with recycled fibres is weaker and easier to tear. In some cases, the more recycled fibres used, the thicker the paper (or paperboard) needs to be in order to compensate for the loss of strength.

Paper already has a good environmental image but efforts are still being put into ways of using other natural raw materials for its manufacture. The potential benefits are even greater if these materials would otherwise be waste products. Two forms of grass stalk are available in large quantities: bagasse from sugar cane and straw from cereal crops. Both can be used to produce very stiff board and paper materials, but even with their almost free availability (and the increasing pressure against burning such waste), the economics are not yet satisfactory. Water hyacinth, a prolific weed which blocks rivers in the tropics and needs to be removed, has also been the subject of experimentation. Finally, at the rarified end, the Japanese Research Institute for Polymers and Textiles has reported experiments in which acetic acid bacteria produce a mass of extremely thin cellulosic fibres from which paper can be formed.

It is a tribute to the cost-containment improvements in the paper industry as well as its inherent economies of scale that even with these alternatives and their perceived potential benefits no widespread substitution of paper has occurred.

Paperboard

Paperboard has dominated the dry foods packaging material market in the twentieth century. Indeed, it is difficult to imagine self-service shopping without paperboard boxes presenting their highly decorated 'billboard' on the shelf.

Paperboard is used in folding cartons for cereals and snack foods, beverage carriers, set-up boxes for confectionery, carded blister packs and gable top milk containers.

The paperboard category includes materials variously termed boxboard, cartonboard, chipboard, containerboard and solid fibreboard. Paperboard is made in the same manner as paper; it can be either a single thick sheet or can be made from multiple layers built up at the web-forming end of the papermaking process. Some of the layers may be made from material recovered from used papers and boards (secondary fibres) or other inexpensive fibres like straw. The top, printable layer is typically made from bleached pulp to give the necessary surface strength and printability.

Solid bleached boxboard is the top quality, made from the purest virgin bleached chemical pulp. Major applications are for high quality foods, cosmetics, electronic components and medical packaging. It is odourless and folds well.

Duplex, or white lined boxboard, has a thin layer of pure bleached white paper over a thicker, partly bleached mixture of mechanical and chemical pulp. This is a very widely used material having an optimum mix of physical, chemical and aesthetic properties. It is suitable for almost all packaging applications, especially food. Triplex grades usually incorporate a core layer of lower grade, often recycled, pulp between two white or other uniformly coloured layers.

Chipboard and newsboard are made from 100% recycled fibre and are the lowest-cost types of paperboard. Colours range from light grey to brown. Chipboard and newsboard are not suitable for printing high quality owing to the short fibres. They are strong and uses include partitions and backings for packages like set-up boxes where appearance and foldability are not critical. In order to improve the scoring and folding, some longer, higher quality fibres can be added. Chipboard can also be covered with a layer of white paper fibres to provide better printability and creasing, and such products are widely used for folding cartons. Lower weights of chipboard are used for the production of spiral wound cylindrical fibre tubes and drums and for the fluting medium for corrugated board.

Solid fibreboard is the thickest type of multi-ply paperboard, made from chipboard, often lined on one or both faces with kraft or a similar paper. The total caliper of the lined board ranges from 0.8 to 2.8 mm. It is used for shipping containers, sometimes in combination with metal or wood-framed edges, where a high puncture resistance is a major consideration. It can also be coated with a water-resistant material like polyethylene for wet uses such as trays for vegetables and fish.

Moulded pulp is not exactly a paperboard form, but is made in a similar way, using primarily recycled fibres laid onto or into moulded shapes. The material is moulded into egg cartons, drinks trays, paper plates, and forms for cushioning or blocking and bracing. In recent years, it has been used increasingly as a biodegradable alternative to expanded polystyrene foam.

Corrugated fibreboard

Nearly 40% of all paper packaging is used in the form of corrugated fibreboard boxes. The structure is an adaptation of the engineering beam principle of two, flat, load-bearing panels separated by a rigid corrugated web. When used to support a stack of boxes, one of the faces is always subject to a tensile stress, and this uses the physical properties of paper in an extremely effective way.

In a well designed box, the most important load-bearing panels have their flutes parallel to the direction of the anticipated load. Since stacking strength is generally most important, most box flutes run vertically. When side-to-side strength is important (in conveyor jam-ups, for example), the flutes may run horizontally.

Corrugated fibreboard is the most common material used for shipping containers, and the regular slotted container (RSC), shown in Figure 2.1 is the most common design. Corrugated fibreboard boxes are well known for their good stacking strength (when dry), easy availability and inexpensive cost. Corrugated fibreboard has also been used to make lightweight pallets and slipsheets (an alternative to pallets).

Corrugated fibreboard is easy to recycle, from both a technical and a logistical point of view. Used boxes are generally discarded

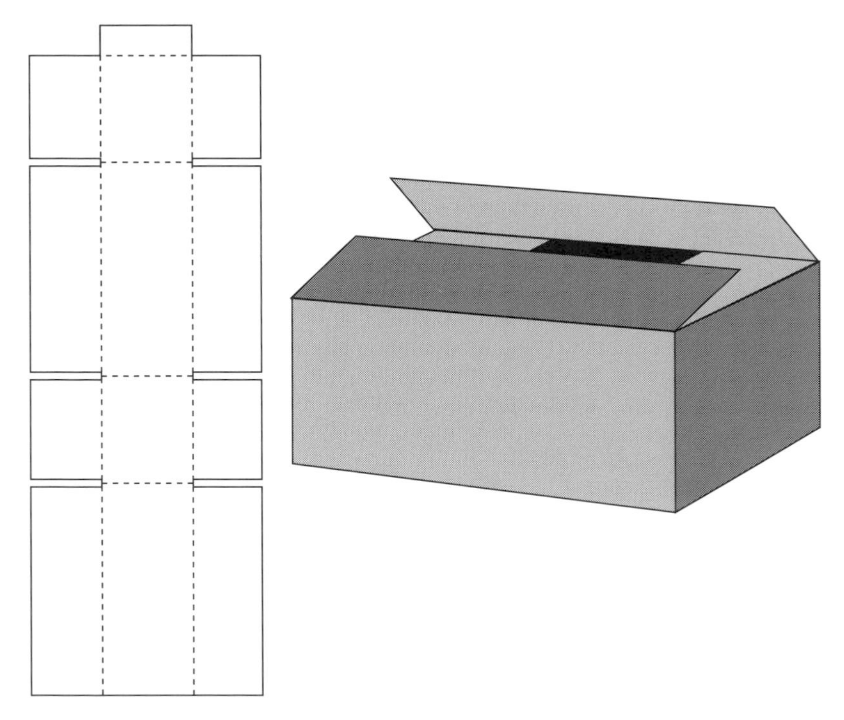

Figure 2.1 Regular slotted container

in large, homogeneous amounts by businesses which have an incentive to reduce disposal cost by recycling. As a result, corrugated board has a very high recycling rate.

The material has been used for almost 100 years, and a series of standard grades has been adopted by most countries. It is categorized in three ways: by the thickness and spacing of the individual flutings of the corrugated fluting medium, by the weight of the facing layers, and by the quality of paper used. The most widely used flute configurations are known simply as A, B, C and E. Dimensions quoted in various sources vary slightly, but British Standard 1133, Section 7 gives the specifications found in Table 2.3.

Table 2.3 Common forms of corrugated fibreboard

Fluting type	Flutes/metre	Flute height (mm)
A	105–125	4.8
B	150–185	2.4
C	120–145	3.6
E	290–320	1.2

Source: BS 1133 Section 7

The first corrugated materials were either coarse A-flute or fine B-flute. The intermediate grade, C-flute has now become the most commonly used type, being a compromise of the best qualities of the other two.

E-flute has very small flutes, and there are even finer grades called microflute, which are used as an alternative to solid fibreboard for display cases and folding cartons. There are other grades with flutes smaller and larger than these standards, but they do not figure significantly in world usage.

Liners to face the board range from 125 to 400 g/m^2 with 150, 200 and 300 grades predominating. The corrugated medium is generally 113 or 127 g/m^2. Heavier weights are generally used for heavier-weight contents, with an aim towards adequate stacking strength.

Corrugated fibreboard boxes are increasingly being used as advertising media, and so higher quality printing is demanded. Three options are possible: direct printing, pre-printed liners and litho lamination.

In direct printing on finished board, the uneven surface due to the flutes limits the quality which is possible, and relatively simple two-colour flexo printing has been the norm. Ink jet printing has gained popularity owing to its print-on-demand capability and, since it is a non-contacting process, the quality is unaffected by variations in rigidity which causes 'washboarding' with conventional printing techniques. The quality of both processes has been much improved in recent years.

Pre-printed liners, high quality flexo printed facing materials, can be built into the corrugated board at the point of manufacture. Developments in paper surfaces, printing presses and the use of polymer flexo plates have made big advances possible in this area.

In litho lamination, printed paper is laminated to the already converted board. Top quality printing is possible including full colour halftone designs.

Direct flexo printing and pre-printed liners are best suited to fairly long production runs since the case design is fixed at the manufacturing stage. Ink jet printing is generally used to add variable text (like lot codes, colours or flavour) to a more generic box, and may be done on the filling line.

Corrugated fibreboard also offers versatility in the number of components which may be combined. The most widely used constructions are single face, single wall, double wall and triple wall, shown in Figure 2.2 and described below.

Single face corrugated fibreboard is a soft material which can be rolled up in one direction and is normally used to provide a protective cushioning for a vulnerable item. Rolled up it can also provide a rigid cylindrical package.

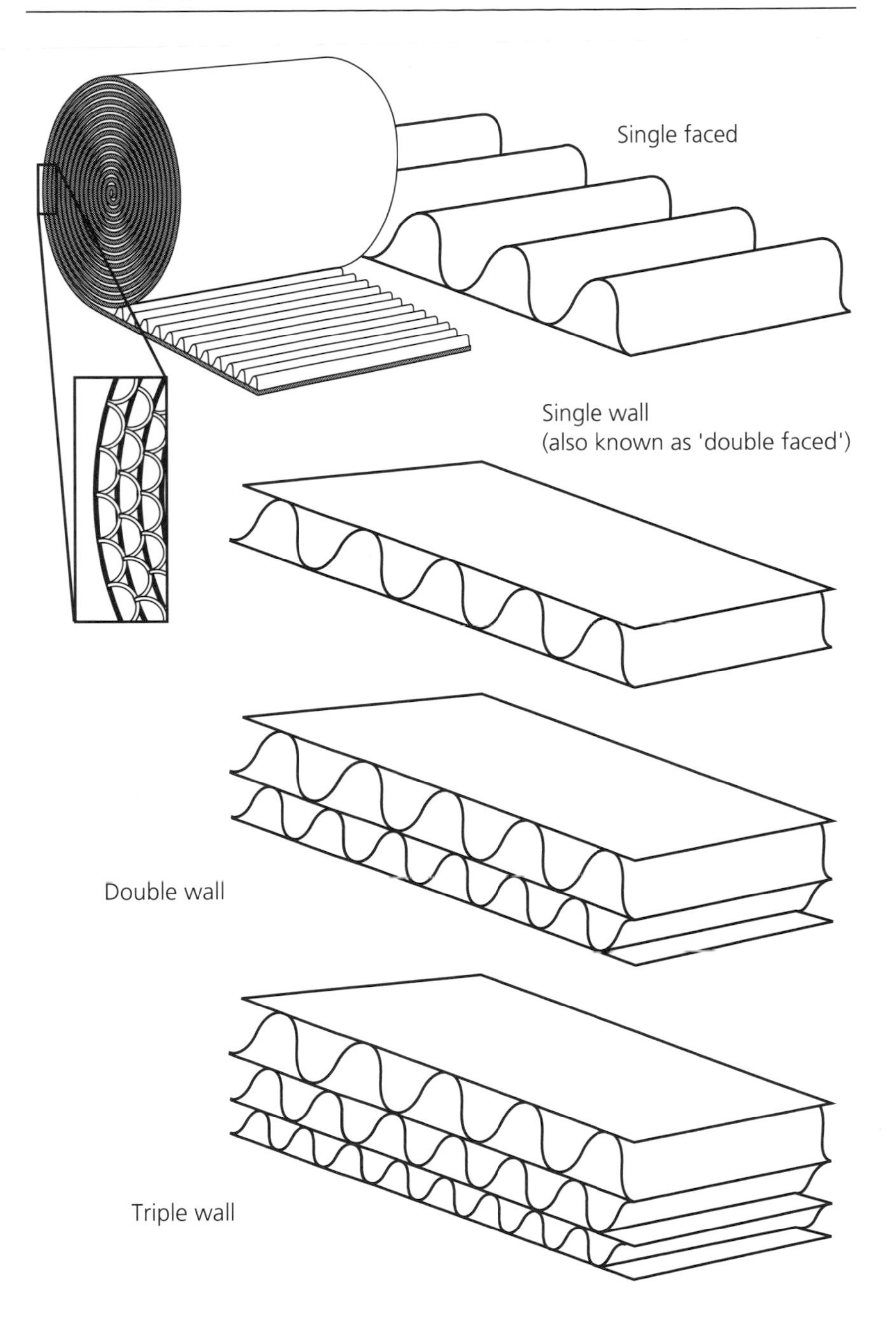

Figure 2.2 Corrugated fibreboard structures

Single wall (a wall is a layer of fluting material faced by two flat sheets) is the most common form, used for cases and trays.

Double wall is another very popular construction, able to incorporate any combination from double A to double E, but AC and AB are among the most popular. Certain constructions are designed to provide high rigidity using an A flute component coupled with a finer grade B or even E, to provide the best surface for printing on the outer facing.

Triple wall can also be made from any combination of flutes. This category includes one of the very heaviest boards available, known as Triwall. With massive facings the material has such a high performance that cases made from it are used in place of wooden crates.

Of course, not all multi-wall boards have such high performance. In countries where wood and paper are scarce, there is a tendency to use multiple walls in an attempt to improve the performance of heavily recycled board, which has short fibres and is therefore weak. In some countries, particularly in the developing world, there is a convention of simply counting up the number of components. Three-, five-, seven-, nine- or even 11-ply corrugated boards may be available, but the individual materials may be of very light weight and offer poor performance.

Corrugated board has an important drawback: it can lose much of its strength (indeed, all of its compression strength) when it is wet. One way to overcome this is to specify that the outer facing is of a wet-strength (resin treated) grade, or a sandwich of kraft/PE/kraft.

Moisture-sensitive starch-based adhesives are normally used in the manufacture of corrugated board, their excellent running properties being needed at the high production speeds involved; but adhesives can also be treated to improve water resistance. Wax dipping or roller coating with wax at the flat sheet stage is also used to confer water resistance, but this is a diminishing practice because of recycling problems and the fact that a wax coating makes subsequent use of adhesives very difficult.

Although very heavy facings can be used, there are limits to the weight of paper which can be put through the heated corrugating rollers to produce the fluting material, 100–130 g/m^2 being the range in normal use. To overcome this restriction, some manufacturers have managed to combine two layers of fluting medium using a very strong adhesive which significantly stiffens the material at the same time, producing a very rigid structure.

There are other variations on the theme. A triple-wall cross-fluted material (e.g., Euroboard's X-ply) has been designed for providing rigidity in both directions and has been used to a small extent for large boxes and pallets. Honeycomb structures incorporate a series of hexagonal cells, rather than corrugated medium, in a thick board (10 mm to 200 mm) that is used for pallets and corner protection where a thick clearance between the outer package and product is needed.

In addition to physical strength, corrugated board provides shock absorbing properties and can even provide some thermal protection owing to the trapped air space. This thermal barrier property is exploited particularly in Japan, where packages to maintain temperatures — either cold or hot — are extremely popular. A corrugated board made from lining materials which are themselves laminated with metallized polyester film provides a reflective surface to trap additional heat or cold.

Corrugated fibreboard shipping containers have become a standard element of most logistical systems. In the USA, transport carriers required their use until transportation was deregulated in 1980. They are easy to purchase and recycle. However, there is increasing competition from plastic alternatives, like shrink-wrap and reusable totes, which are lower cost in some situations. There is also a trend towards the use of point-of-purchase and pallet-load displays which may use corrugated fibreboard in more innovative ways with fewer (or no) traditional shipping containers.

Wood

Only in Japan is wood used to any significant extent for retail packs today, and even the traditional cedar wood cigar boxes are being replaced by plastics. Most of the applications in the western world are for heavy-duty transit and industrial packaging, where the high stiffness, low weight and versatile construction options of wood can be best employed. Wooden barrels are still used for wine aging, but are now rarely used for transit packaging.

Pallets are the primary application for wood in packaging. Despite the increasing use of alternatives made from plastic or fibreboard, the wooden pallet has not been matched for versatility, reusability and repairability. Wooden pallets are ubiquitous, an important component of modern distribution in which mechanical handling predominates.

Choice of wood species has a great impact on cost and durability. The denser and stiffer the wood, the greater its durability and cost; hardwoods (like oak) are the most durable and costly.

Despite the undoubted advantages which wood can offer, it has also some serious limitations, and attempts to resolve these have accounted for much research effort in recent decades. These limitations are moisture content and directional weakness.

Timber used for packaging should be well seasoned, either by air drying or kiln drying. Drying is an essential preparation for some wooden packages because it reduces shrinkage, protects from microorganisms, reduces weight and increases rigidity. Moisture content is of particular importance if, for instance, a wooden pallet is sealed inside a moisture barrier wrapping, for products like machinery. The pallet may hold up to its own weight in water, enough to swamp the capacity of most desiccant packs (described in Chapter 14) which may be included for the purpose of absorbing moisture. Impregnating wood with resins can reduce the tendency to absorb moisture, but the rate of resin penetration into large pieces of solid wood is slow, and so composite structures are generally used where moisture sensitivity is a critical element.

The directional strength (and conversely in the opposite direction, weakness) of thin sheets has for years been compensated for by cross-laminating in plywood structures. Plywood has extremely high stiffness and puncture resistance, is lightweight, and can be joined by all methods, from adhesives to plastic clips and nails. Plywood is available in many grades and thicknesses and is relatively inexpensive.

Reusable plywood boxes are frequently made using patented metal or plastic joining systems. Their ability to be collapsed and delivered flat gives these cases a great advantage over traditional rigid wooden versions. Heavier crates for machinery, etc., often incorporate plywood panels framed in solid wood. These offer the advantage to end recipients of being able to re-use the wood — an especial benefit when shipping to countries where wood is a scarce resource.

Hardboard, made from wood chips and fibres, is another wood-derived material which has some distinct advantages. Being made from a much larger proportion of the tree, and on a continuous process, it occupies a position on the cost scale between timber and plywood on the one hand and corrugated board on the other. It is strong, but brittle. Since some binding component must be incorporated to hold the particles together, it can double as a moisture-resistant and even decorative finish. A more recent approach involves the blending in of elastomeric materials such as rubber to increase the material's toughness. Hardboard is available in various grades and in thicknesses from 2 to 12 mm, with various surface finishes and textures. Panels can be used as direct substitutes for plywood, although they may need to be thicker. The material's high rigidity is exploited by incorporating it as stiffening panels in combination with corrugated or flexible materials.

Another form in which wood is used for packaging is a formable reconstituted sheet material produced from woodchips and shavings. Since this type of material is blended, conditioned, mixed with a binding agent and compressed under heat, it is possible to produce shaped items in a single process. Some nestable pallets are made in this manner.

Other wood-like fibres can be combined with resins used to produce rigid panels. These range from residues of sugar cane to straw and plant stems. The natural bonding ability of cellulose fibres, which is important for papermaking, cannot be relied upon to form a strong enough material with such fibres so organic resins are usually incorporated. An example is called Compak Board, produced on a hot press operation from chopped straw — a material which is generally believed to have no economic value. Other fibrous materials like rice husks, rape stems or woodchips can also be incorporated, and the properties can be varied according to the length and mix of fibres, proportion of resin and density of compaction.

Cork

Cork closures, made from the bark of the cork oak, have a history dating from the Roman Empire when they were used as bungs for wooden barrels and (presumably) also for closing glass and ceramic bottles. After the Industrial Revolution their use as bottle closures grew, as they represented an improvement over competing materials like glass, clay, and fibres coated with wax or rosin. By the mid-nineteenth century, cork was the predominant closure material for bottles.

Cork has since been replaced in most bottle closure applications by caps and closures made from metal and plastic, although it is sometimes used as a liner material for caps.

Its most important use today is for closing wine bottles. Cork closures are valued because they give a good positive seal for glass bottles. Most corks are made from trees in Spain and Portugal. Cork suitable for wine stoppers cannot be obtained until the tree is 30 years old. After air drying, the cork is boiled to shrink it and destroy insects and mould.

In recent years, wine bottle closures have increasingly been made from plastic and composite materials, owing to insufficient supply of cork. There is also a quality problem with corks, since many are infected with a particular species of bacterium which changes the taste of the wine (which is said to be 'corked'). There is a great deal of controversy about how to detect and eliminate this problem, since it is not predictable or consistent. Despite these problems, cork closures are still favoured for fine wines for traditional, rather than technical, reasons.

3
Glass

Glass is another very old material, discovered, it is said, by Phoenicians who lit a fire on the beach and later found that the sand had melted. However, its widespread use for packaging goes back only about 200 years, for commercial packages such as wine bottles.

Its value is clear: glass is strong, durable and transparent. It is chemically inert and is an absolute barrier to moisture and gas. Glass bottles can be sterilized and will withstand high food-processing temperatures. It is easy to recycle. Glass bottles convey an image of quality, and are used for upscale niche food products like wine, premium beer, perfume and condiment sauces.

Glass is still an important packaging material, even though it has lost much of its market share to plastics and aluminium. The drawbacks of glass are its weight and breakability.

The primary ingredients are inexpensive, and they have not changed over the centuries: the main constituent of glass is silica (sand) with smaller amounts of soda and lime. There has been a constant stream of technical improvements to the materials and to the processing. Modern glass contains a number of other minor ingredients to improve its melting ability, strength and appearance. A typical bottle grade of glass may include calcium oxide (10–12%), sodium oxide (12–15%), magnesium oxide (0.5–3.0%), alumina (1.5–2.0%), and trace amounts of iron oxide and sulphur trioxide.

The ingredients are crushed, blended with about 20% of scrap glass (called cullet) and fired in a furnace at about 1300°C. Higher levels of cullet can be used, and the growth of recycling is making this possible. Coloured grades are made by the addition of traces of chrome oxide (green), cobalt oxide (blue) and iron plus sulphur (brown).

An invisible but important ingredient is energy, and the large amount needed has a big influence on the production economics. Several techniques have been developed to reduce energy use, including pre-blending and pre-heating of the raw material, as well as better monitoring and control of the processes.

Greater use of cullet also reduces the energy needed, because it is considerably easier to melt. The glass industry has put a great deal of effort and resources into encouraging the recycling of used glass bottles for reasons related to economics, the environment and public relations.

Most glass bottles are formed in a two-step blow and blow process. A gob of hot glass is dropped into the mould and pushed into the neck, called the finish. (In hand-blown bottles, the 'finish' was the last part formed; in modern bottle-making machines, it is the first.) Air is blown into the finish in a two-stage process, first blowing a test-tube shaped parison, and next reheating and blowing into the parison until the walls conform to the mould cavity. A press and blow process is used for wide mouthed jars, where the parison is formed by a plunger. After forming, annealing in an oven is required to relieve stresses as a result of moulding.

Vials, ampules, pipettes and very small bottles are made from glass tubing which is made by forcing molten glass through a die. The tubing is trimmed and glazed in a separate operation.

Lightweight and strength improvements

The economics of glass containers are dependent on weight. Weight is related to the amount of raw materials needed as well as to the distribution economics, since freight costs are directly dependent on weight.

Therefore, lightweighting, without sacrificing strength, has long been a goal in glassmaking. The UK's returnable one-pint milk bottle has been progressively reduced in weight from 600 g in 1920 to 225 g in 1998.

It is well known that glass is an enormously strong material even in very thin sections (one has only to think of glass fibre reinforcing filaments and the common electric light bulb), so it is theoretically possible to make very thin containers in glass.

The weakness of glass is that it is brittle and tends to have its stresses locked into the surface layer. If the surface is damaged at all, the material can be easily broken — which is how a glazier can score and then snap a thick sheet of glass between his or her fingers. Likewise, a scratched glass bottle filled with pressurized liquid can become a dangerous bomb when dropped.

Much research has therefore gone into the four main routes to minimize this surface damage factor in thinner containers:

- reduce internal stresses;
- minimize surface blemishes;
- coat the surface; and
- use protective labels.

The internal stresses can be reduced by design changes which result in better distribution of glass and improved annealing control. Since glass is a supercooled liquid, it can cool at differing rates according to air currents, varying wall thickness, conductive effects from hot support surfaces and the proximity of other hot containers.

For example, high stress concentrations in sharp transition regions tend to reduce a bottle's strength: a bottle with a smooth vertical profile is stronger than one with sharp transitions.

Similarly, a cylindrical bottle is usually stronger than a rectangular one. Computer modelling shows that a balanced design of the shoulder, heel and bottom regions can contribute to strength. Better understanding of these complex effects has made it possible to reduce bottle weights efficiently.

External surface blemishes can be reduced by careful manufacture and handling of the moulds, and by equally careful handling of the glass parison and bottle during manufacture and distribution. This is an important factor to consider in the design of multi-trip bottles. An analysis of the points of contact with shipping containers, guide rails, base supports, closure and labelling equipment makes it possible to build in 'wear absorbing' zones or strengthened sections. An example which allows for absorbing wear is to dimple the contact surfaces (usually around the sides and base); the high spots will be rubbed away in a controlled manner with minimum effect on the overall strength.

There are two kinds of surface coating available, depending on whether the coating is applied on the hot end or the cold end of the bottle-making process. Hot-end treatments toughen the outer skin. Cold-end surface treatments lubricate the external surface. Both types reduce the possible damaging effects of the glass-to-glass contacts which are unavoidable during normal filling and use.

A combination coating of tin or titanium oxide followed by PE is the most usual scratch-resistant coating. Alternative cold-end coatings are based on polyurethane, epoxy, acrylic, stearates, oleic acid, silicones and waxes.[1] There can be bonding problems which result in the coating coming into contact with the food.

Some of the cold-end treatments are removed by the alkaline wash water used in returnable systems (for example, as used for beer and milk in the UK). Some returnable systems have experimented with recoating the bottles after washing.

Other creative surface coating-type modifications include: fusing a skin layer of a different kind of glass onto the bottle exterior, modifying the surface chemistry of the glass, and other experimental coatings.[2] Some surface treatments and coatings can also be used to change the colour or surface texture, a technique which is used extensively in Japan. One example is a double coating of plastics (tradename Mul-t-Cote from Star Chemical in Japan): styrene butadiene rubber followed by a thinner layer of high modulus polyurethane which may be coloured.

There are a number of protective sleeve label solutions to the damage problem. The most common is a foamed polystyrene (PS) shrinkable sleeve about 1 mm thick, covering the stress areas from the shoulder to under the base rim (tradename Plastishield), which can be printed. In addition to surface protection and the ability to lightweight the glass bottle, other benefits include reduced noise during filling and a better and safer grip for the consumer with some degree of thermal protection.

Polyvinyl chloride (PVC) reverse-printed shrink-sleeves have been used in Japan for 20 years, and more recently have expanded into Europe and the UK. The label is usually printed, but can also include an overall colour to convert a clear glass bottle to an apparently coloured one. Recently, selective metallization and foil lamination have been added to the options.

In addition to protecting the container from surface damage by glass-to-glass contact, an important function of the double-coating and the protective label solutions is to prevent dangerous scattering of glass shards if a pressurized bottle is dropped.

New developments

Research in coating and strengthening techniques continues, but there is more emphasis now in marketing. The glass bottle industry has survived, but only barely, the competition from plastic and metal containers. Tinplate and aluminium cans predominate in the beer market, and plastic has replaced glass for soft drinks, water and many other food products.

The glass industry is concentrating on the market niches where it provides a competitive advantage, like premium food products and for refillable bottle applications. Glass bottles are still the best choice for refillable bottles, and can be easily recycled.

The industry has put a great deal of effort and resources into encouraging the recycling and re-use of glass bottles. This environmental factor could be the most significant for the future of glass packaging, since some countries are legislating the use of refillable and recyclable bottles.

Other new developments are related to improving plastics with a material chemically similar to glass. Silicon oxide can be vapour deposited in an ultra thin layer on the surface of plastic materials, resulting in a flexible material with nearly the properties of glass. This is described in more detail in Chapter 13.

Similar in some respects, but different in technology, is the research into novel ways to produce continuous layers of glass using a cold precipitation process. A solution of tetraethoxysilane and water is heated to produce a glassy layer in situ. Initial applications have concentrated on very thin layers for microchips but it could have packaging applications in the future.

4
Metals

The most important metals used in packaging are steel, tin and aluminium. Tinplated steel food cans and aluminium beverage containers are the most prevalent applications.

Other metal packaging applications include steel drums and pails, closures, strapping, trays, and as a thin foil barrier layer in laminated materials. In these applications, metal is coming under increasing pressure from plastics materials which have inherent advantages, since it is easier to produce complex designs like closures from plastics, and plastics are self-colouring and heat sealable.

However, metals have performance advantages of their own. Of all packaging materials, metals have the highest absolute performance in heat tolerance, physical strength and durability, barrier and absence of flavour or odour, stiffness and deadfold. Cans have the added advantage over glass bottles of easy mass handling without breakage, lightfastness, and the ability to be produced and filled at higher speeds.

For different packaging applications, the importance of each of these properties varies, and hence the rate of substitution differs similarly. For a number of products, like beer and canned food, metal remains the material of choice in most countries.

Cans

Tinplated steel was originally used for tea and tobacco boxes and canisters. Tinplate sanitary food cans were developed in the early 1800s, shortly after Appert's invention of 'canning', the retort food preservation method (which originally utilized glass bottles).

Cans can also be made from tin-free steel (black plate and chromed plate) and aluminium. In some countries, most beer and soft drinks cans are made from aluminium.

Tinplated steel and tin-free grades

Steel is made from iron and a small amount of carbon, which is why tinplate and aluminium cans can be separated for recycling with a magnet.

Tinplated steel can stock is made by hot rolling and tempering the steel sheet to a standard thickness. Tempering determines hardness, and several grades of temper are available. Harder grades are used for can ends; more ductile grades are used to create shapes, like drawn cans (described shortly).

Bare steel rusts easily and unprotected black plate can be used only for non-corrosive products like waxes and oils. More commonly, tinplate is electrolytically applied in a very thin coating. After trimming, most of the coiled reel stock is coated with an organic lacquer to further protect the steel against corrosion. The tinplate and lacquer also protect the iron from being dissolved in food products, which would leave a bad taste.

Foods that are more highly acid require a heavier tinplate or lacquer coating (or both). (The acidity level of a liquid can be meaured by its pH value; the lower the pH value, the higher the acidity.) Different lacquers have been developed for the packaging of various foods which vary in acidity. The thickness of the tinplate coating may not be the same on both sides. The surface facing outwards is often thinner because it is exposed only to ambient humidity and not to the contents.

Lightweighting has taken place steadily in the can industry. Since 1945 the weight of steel in a processed food can has been reduced by 35% and the tin by 80%. In modern cans, the tin layer accounts for only 0.4–0.5% by weight.

Developments in tinplate over recent years include the use of water-based lacquers for environmental reasons and greater use of lacquered tin-free steel.

Tin-free steel or electrolytic chrome-coated steel was developed in Japan during the 1960s at a time when tin prices looked set to rise sharply. The chromium metal plus chromium oxide coating is much thinner than comparable tinplate. It produces a bright metallic finish but does not give the same degree of protection against corrosion as does tin, so it is essential that it be lacquered. It is necessary to remove the chrome in order to weld the material, which is often used for can ends where weldability is not a concern. Similarly, a new can-making process (Toyo Seikan's TULC and British Steel's RBS trademarks) laminates polyester (PET) film on both sides of a tin-free steel substrate.[3]

Two- and three-piece can-making

More significant than the material developments have been those associated with can manufacturing. By far the most significant of these, and one which has found its greatest application in the carbonated beverage and beer sector, is the two-piece can.

For most of their history, tinplate cans were made in three pieces: two ends and a body blank. Today the side seam is usually induction welded, replacing the original lead-based soldering. By comparison, welding saves material (less overlap is required) and produces a stronger seam, which is important for pressurized containers like aerosols. Newer welding techniques include the use of electrical induction and lasers, which permit an increase in production speeds. For some products, the seam may be cemented rather than welded or soldered.

To complete the three-piece can, the two ends are seamed onto welded body cylinder, one by the can-maker and the second after

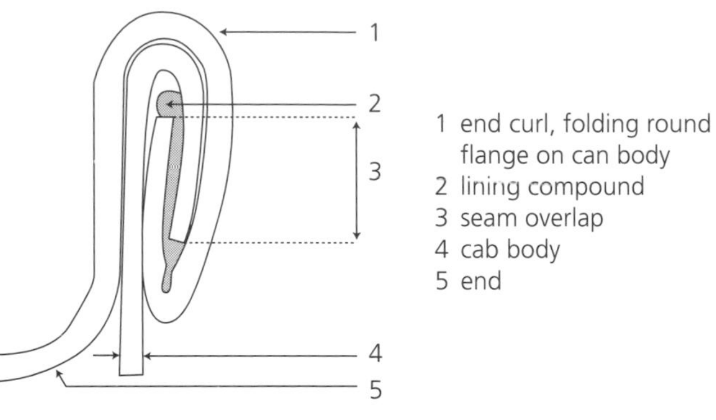

Figure 4.1 Double seam

filling. The double seam includes a sealing compound, shown in Figure 4.1, which is critical in forming the can's strong airtight seal.

On the other hand, the two-piece can is made by stamping the metal in the shape of a cup, from a round disc blank, using very high pressures and progressive dies. The filler applies the double-seamed (top) can end.

Three methods are used for making two-piece cans, depending on the depth of the cup: drawn, draw–redrawn, and drawn and wall-ironed. Shallow cans, commonly used for products like tunafish, can be drawn directly by stamping a die into a metal blank.

Cans for carbonated beverages are made by the drawn and wall-ironed (known as D&I in the USA and DWI in Europe) technique, which involves stamping out a deep cup from a sheet which has the same base as its intended final diameter, by stretching the walls to the desired height.

The walls of DWI cans (especially aluminium cans) are very thin and if it were not for the interior pressure of the carbonated contents (soft drinks are about 3.5 atmospheres or 50 lb/in^2), these cans would collapse. Likewise, DWI cans cannot be used for retorted food because they cannot withstand the vacuum. Non-pressurized processed foods have been packaged in aluminium

DWI cans under nitrogen pressure, which extends shelf-life and helps to support the can walls.[4] Aluminium cans are discussed in more depth shortly.

Draw–redraw (DRD), a more versatile approach, uses more metal since an oversized cup is first stamped out and this is then pressed by stages into deeper, narrower stamping dies. As a result, the walls are thicker and more uniform than DWI cans. DRD cans are made of electrolytic chrome-coated steel coated with enamel and can be used for food.

Two-piece can technology results in considerable savings in material because the walls are thinner. This saving is important, because (as shown in Table 1.3 above), metal accounts for most (about 75%) of the cost of the can. However, round two-piece blanks leave a significant amount of waste, whereas three-piece body blanks are square and result in no waste (but the ends do, hence one of the incentives to use smaller ends).

The most common non-beverage product sectors in which two-piece cans are used in the UK and USA are pet foods and baby foods. There is a large investment needed for high capacity production of two-piece containers, and they are primarily available in standard sizes.

A further example of materials saving is in the reduction in diameter of the top end. The technique of necking-in was developed initially for three-piece cans to eliminate the projecting beads at the top and bottom which gave rise to the phenomenon known as 'rim riding' in transit, and resulted in scuffing, dents and, sometimes, leakage.

With the advent of the two-piece can it was realized that necking-in could be taken further for economic reasons. Making the thicker can end smaller results in the use of less material, less seaming time, and less sloshing on liquid filling lines. The high-speed production economics of canning are such that small per-package savings add up quickly.

Most aerosol cans are made from tinplate, in a technique similar to the three-piece can process. The base and dome with a 1″ aperture, are applied to the body with double seams and shipped to the filler. Some aerosols are made from aluminium in a two-piece DWI process.

The aerosol can is partly filled with the product, a valve is inserted into the top opening and hermetically sealed to the can by a crimping process. Then the aerosol can is gassed with a propellant that can be easily compressed and liquefied at the operating pressures in an aerosol system; most use hydrocarbons. Chlorofluorocarbons (CFCs), the first propellants, have been discontinued because of their implication in depletion of the atmospheric ozone layer. In the meantime, several hydrofluorocarbons (HFCs) with less environmental impact, have been growing in use.

Aluminium cans

The two-piece can-making process and the necked-in end are particularly important innovations for aluminium can economics. Aluminium is one of the most expensive packaging materials available, and so great efforts have gone into such material reduction strategies.

Aluminium's high value is also the reason for its high recycling rate. Scrap aluminium has a high value, and aluminium manufacturing is less costly when using recycled stock. The concentrated industry of large aluminium can manufacturers has facilitated recycling programmes.

Over 98% of aluminium cans are DWI and are used for beer and soft drinks. Aluminium cans are completely substitutable for tinplate in the beverage sector, and the proportion of beer and soft drinks packed in each varies in different parts of the world. The choice depends on relative economics, energy cost and available technology. For example, the USA, Italy, Austria, Sweden and Greece use almost 100% aluminium cans, and Germany, Belgium, the Netherlands, Spain and France use more tinplate.

So far the penetration of aluminium cans in the processed food sector has been low. The thin walls cannot withstand the vacuum that is created by retorting.

Aluminium is more ductile than tinplate, so it can be rolled or drawn into thinner sections. Its ductility allows a greater degree of necking-in than does steel. A Japanese process makes an eight-step necked-in can that looks more like a bottle, with a diameter reduction of 20%, but too much necking can increase production costs. In most applications, three- to four-step necking-in seems to be about the optimum.

Easy-open can ends are easier to make from aluminium than from tinplate, and most aluminium cans have them. Tinplate is harder and less ductile than aluminium, and quickly blunts the punching knives. With tinplate, there is more of a chance of either cutting too deeply, ruining the can's hermetic seal, or not cutting deeply enough, which makes it difficult for the consumer to open the can.

Opening food cans by full aperture easy-open ends poses a danger because of sharp edges, but consumers are sometimes surprised to find that easy-open cans are also sharp. A further problem is that the pre-scored cut has to be a minimum distance from the seam, projecting from the can wall. There are some designs that mitigate this problem, forming a folded guard for the sharp cut edges.

The success of the two-piece aluminium beverage can is partly owing to the pull-tab opening device. Generally, easy-open aluminium can ends have a high magnesium content. To reduce litter and increase safety, in most designs the tab remains attached to the can end.

New can developments

Owing to the ductility of aluminium and continuing improvements in this by varying the metallic make-up of the different alloys, as well as better mechanical engineering in the forming process, a continuous programme of weight reduction has been possible. The metal content of a 33 cl aluminium beverage can has fallen by 25% over the past two decades, but the scope for further material reductions is now very small.

Therefore, aluminium can-makers have turned to developing cans with more aesthetic appeal and novelty. The current trend is towards more novel can shapes. Coca-Cola has developed a single-serving contoured aluminium can, to match its traditional, instantly recognizable, curvy glass bottle silhouette, following the successful introduction of similarly shaped PET bottles.[5] A new Heineken/Metal Box can reflects the curvy shape of a traditional British pint beer glass.[6]

There have been other novelty cans developed, especially in Japan. Many include easy-open and reclose devices. There are devices which reproduce the foaming head on draught beer when the can is emptied into a glass.

Self-heating and self-chilling cans have been invented, making use of exothermic and endothermic chemical reactions inside separate sealed chambers. The self-heating can uses the reaction between quicklime and water. Self-cooling cans use sodium nitrate and water or liquid carbon dioxide.[7]

Aluminium foil and trays

Aluminium foil is produced by rolling aluminium through heated rollers or by casting and cold rolling. Packaging end uses range from semi-rigid oven-heatable trays to laminations with plastic.

Aluminium foil provides the best flexible barrier available; it is impermeable to water, gas and aromas. It is lightweight and resists most chemicals and oils. It is stable in hot and cold temperatures; food can be frozen and then cooked in it. It is formable and can hold a 'dead fold,' although creases can create pinholes.

Aluminium foil was first used for wrapping candy. It still has many wrapping uses, but is now usually laminated to another substrate like waxed paper or plastic. This enables a very thin layer of aluminium to be used as a barrier while minimizing pinhole problems. Foil laminates are used as wrapping for gum and cigarettes, cereal box liners, and lidding for cups of dairy products and drinks. Aluminium foil is used in laminated toothpaste tubes.

One of the most important applications is the use of aluminium foil laminate material in form–fill–seal operations for liquid heat-processed food products. Retort pouches are typically a lamination of PET (on the outside, for strength), aluminium (for its barrier properties), and PP as a sealing layer. Aseptically packed liquid cartons often include a barrier layer of aluminium along with paper (for printing) and PE for sealing.

Pressed aluminium trays, made from thick foil, were originally developed to supply bakers with disposable baking pans and pie plates. They became popular as the plate for frozen 'TV dinners' in the 1960s, because a meal for one could go from the freezer to the oven to be eaten (presumably in front of the television) in the same tray. Aluminium trays have also been used extensively for chilled foods in catering, institutional cooking and takeaway hot meals.

With the advent of the microwave oven, demand for convenient prepared meals has grown dramatically. Foods prepared specifically for the microwave are growing, and trays for this market are produced from coated board moulded pulp, plastics and aluminium.

There was some controversy about the suitability of aluminium for use in early microwave ovens, arising from the fact that arcing occurred when metal objects were placed in the oven during operation, sometimes disabling the unit. This was largely corrected when the electronics were later improved. There have also been questions about whether aluminium shields some areas during reheating; a 1984 study found no significant difference between the temperature of food microwaved in aluminium versus other microwavable trays, but 1988 and 1989 studies found variations in heating effectiveness based on different ovens, products and packages, as well as between different areas within a single tray.

There have been other metal foils developed and used in packaging, although they are now rare because plastics and aluminium are more cost effective. Tin foil (made from an alloy of tin and either lead, antimony, zinc or copper) was used for wrapping chocolates as long ago as the 1840s. Lead foil was used to line tea boxes in the 1800s and was used in cigarette packaging as late as the 1930s. Lead seals were used for wine bottles until the 1980s.

Iron foil has been produced commercially for many years and its possible uses in packaging have been re-examined periodically. It has been laminated to plastic film in Japan and Europe and used to make portion packs for jelly. The material can be produced by hot rolling, direct formation from the melt via a slit die, by sintered powder compression or by electrolytic deposition. Its main drawback is that the very bright and reactive surface rusts easily and so needs immediate protection.

Steel drums and pails

Heavy-gauge steel is used to make drums and pails for dangerous goods and other liquids shipped in semi-bulk quantities. These were introduced in the early 1900s as substitutes for wooden barrels, especially for petroleum products. Barrels, drums and pails are strong — a cylinder is the strongest shape — and can be handled easily by a single person (rolling it), although most are now palletized and handled mechanically.

Most drums and pails (pails are smaller than drums) are made from steel treated to resist rusting. Coatings are applied to the inside, phenolics to protect against acids, and epoxies to protect against alkalis. The exterior is painted to provide additional protection from rust and for decoration.

There are two styles: tight-head, in which the top of the drum is fixed and has a threaded opening which is used for filling; and open-head, in which the removable top is secured by using a separate closing ring. The containers are seamed in a manner similar to that used in can manufacturing.

Endnotes for Part 2

1 Johansen, R 'Favorable conditions for plastic coated glass bottles?' *In-Pak* Vol 14 Nos 6–7 (1993) p 25

2 Doyle, PJ 'Recent developments in the production of stronger glass containers' *Packaging Technology and Science* Vol 1 No 1 (1988) pp 47–53

3 'Defending the steel can in Japan' *Can Technology International* Vol 3 No 8 (1996) pp 22–25; 'Laminated cans for new markets' *Canner* (June 1997) p 6

4 'Nitrogen stabilizes thin-walled containers' *Packaging Report* No 9 (Sept 1990) p 30

5 Lindsay, D 'Shaped to sell' *Beverage World* Vol 116 No 1635 (15 Mar 1997) pp 91–2, 94

6 Brown, M 'Heineken shapes up' *Canmaker* Vol 19 (Nov 1997) pp 19–20

7 Newman, P 'Just one look' *Canner* (August 1995) pp 32–6

Part 3 Synthetic materials

5
Introduction to plastics

The word plastic describes materials which can be made soft and malleable, capable of being moulded or shaped, which are then fixed by heating, chemical reaction, or cooling. As such, most materials — including clay, metals and glass — can strictly be described as 'plastic' materials. The term is now used, however, to describe synthetic materials with the ability to be formed into useful shapes by means of heat — a shortened version of the word thermoplastic.

The first of what we now call plastics were developed by Mr Parkes over 100 years ago during the search for substitutes for natural decorative materials such as ivory, ebony and tortoiseshell. The early plastics were mainly of the type known as thermoset plastics, i.e., those which once moulded into the final form and set by heat cannot be subsequently softened. During the 1930s to 1950s these became the major types of plastics; Bakelite is the best known name. These materials, based on phenol formaldehyde, urea formaldehyde or melamine formaldehyde, found applications in packaging mainly as rigid closures and fittings, since their brittleness and forming limitations made them unsuitable for most containers.

Only four thermosets are still used to any extent in packaging. Phenol formaldehyde (PF) and urea formaldehyde (UF) are used mainly for bottle closures. UF is resistant to oils and solvents and is used in the cosmetics industry; PF is used for pharmaceutical closures because it is more resistant to water. Both PF and UF closures have been largely displaced by polypropylene. Glass-reinforced polyester, another thermoset, has been used for storage tanks and large transit containers. Polyurethane, used in foam cushioning, is also a thermoset.

Regenerated cellulose film (also known as cellophane) is also one of the oldest synthetic materials, but does not qualify as a plastic material on the basis of either its origin or its nature, as described in Chapter 11. It is derived from natural cellulose (wood, primarily), and it cannot be moulded since it can be manufactured only as a thin sheet. Since it is natural cellulose, it does not soften with heat, but chars like paper (a flame is the easiest way to test for it), rather than melting like plastic. Its early name was 'transparent paper'.

Thermoplastic (as opposed to thermoset) materials were developed during the 1930s. Celluloid (cellulose nitrate) and cellulose acetate, derived from natural cellulose materials, and Perspex (the tradename for polymethylmethacrylate) are examples, but their use in packaging was initially very restricted.

From the 1950s on, thermoplastic materials became more widely available and the chemical term polymer was adopted. The word means literally many parts, referring to the linking together of monomers, smaller molecules, in chains.

The arrangement of molecule chains affects the properties of the material. If the chains are randomly arranged, the plastic is called amorphous, and is low density and stretchy. If they are lined up parallel, the plastic is called crystalline, and is usually stiffer and of higher density. Polyethylene, for example, can range from high to very low density according to different arrangements of the same molecules.

Thermoplastics are by far the dominant form of all plastics in use today. They are all based on organic chemistry (because their long chains of carbon atoms can build up into giant molecules) and at present they are virtually all derived from petrochemical feedstocks — mainly crude oil. They can in fact be produced from other organic sources including vegetable materials but current economic factors make these of marginal interest, although in the long term they will probably become significant.

Most of the major polymers are derived from simple gases like ethylene and propylene, the molecules of which join together to form polyethylene and polypropylene respectively. However, the technology to split the petroleum gases into pure monomers (like ethylene and propylene), and to induce these to polymerize, under the influence of pressure and catalysts, into solid materials (like polyethylene and polypropylene) is complex and expensive. Only the ready availability of the raw materials and the economy of large-scale processing make the materials so low in cost.

There are about a dozen plastics materials commonly used in packaging, producing a spectrum of properties to match most needs. The principal types of thermoplastic used for packaging are polyethylene (PE), polypropylene (PP), polyvinyl chloride (PVC), polystyrene (PS), polyester (PET and PEN) and polyamide (nylon). Chapters 6–10 discuss these commodity packaging plastics and related polymers.

Table 5.1 estimates worldwide plastics consumption at 151 million tons by the year 2000. The materials with the highest production are low density polyethylene and polyvinyl chloride. Polyester is the fastest growing. Table 5.2 shows that the polyethylenes constitute the largest proportion of polymers used for European packaging.

In addition, plastics can be combined in many different types of structures to provide levels of performance not available from any single material. Chapter 13 explores composite flexible materials and their unique properties. The plastics in use today have been around for 10–50 years and can be combined to meet most of the needs of packaging.

The main limitations of plastics are in barrier performance and heat tolerance. A polymer offering improved performance in these particular respects at an acceptable cost could end the search for the 'ideal' single plastic material.

Table 5.1 World plastics apparent consumption, 1994–2000 (million tons, rounded)

	1994	1995	2000	AAGR % 1995–2000
Production	121	127	152	3.6
Consumption				
LDPE	22	23	30	5.5
HDPE	15	16	20	4.6
PP	18	19	26	6.5
PS	10	10	13	5.4
PVC	20	22	28	4.9
ABS	3	3	4	5.9
PET	0.8	0.8	1.2	8.4
Others	28	27	29	
All plastics	117	121	151	4.5

Source: Business Communication Company (August 1997). ©1997 The Dialog Corporation plc

Table 5.2 Polymer types used in packaging, Europe only, 1989–95

	1989			1995		
	Total use '000 tonnes	Packaging use '000 tonnes	Share %	Total use '000 tonnes	Packaging use '000 tonnes	Share %
LDPE and LLDPE	5 250	3804	72.5	5 825	4 340	74.5
HDPE	2 720	1496	55.0	3 597	2 295	63.8
PS and EPS	1 720	1000	58.1	2 352	957	40.7
PP	3 350	958	28.6	3 692	1 677	45.4
PVC	4 880	889	18.2	5 401	860	15.9
PET		299		800	672	84.0
Others	1 707	103	6.0	1 787	186	10.1
Total	**19 627**	**8549**	**43.6**	**23 454**	**10 987**	**46.8**

Source: APME

The field of engineering plastics actively continues research in applications with higher performance demands. Research into these can sometimes lead to the development of new polymers appropriate for packaging uses, or they can lead to new production methods which make some of the hitherto expensive engineering materials available at much lower costs, and hence able to be considered for wider applications such as packaging. Chapter 12 describes packaging applications for some high performance materials.

Improvements in performance or usefulness do not necessarily require the development of new polymers. There are many variations and permutations of the current plastics which can extend the range of useful applications.

Performance-enhancing developments include process modifications such as the use of different catalysts, combining two or more monomers (to form copolymers and terpolymers), and the blending of different materials. There is a long list of possible permutations, combining anything from two to ten different materials (plastics and non-plastics) in some form of multi-layer or blend structure, thus allowing the best properties of each to be employed in the most economical way. Chapter 13 discusses multi-layer structures, blended plastics and additives.

Additives can be used to add specific properties when needed. Some examples include antifogging agents, antioxidants, antistats, colours, flame retardants, foaming agents, lubricants, mould release agents, plasticizer, heat and ultraviolet stabilizers, and surface reactive agents. Fillers (such as glass and minerals) and reinforcing materials (like fibres) are also used to impart specific properties.

It is also possible to modify the properties of a basic material after it has been produced — either in polymer form or in its converted form as a film or container. For example, irradiation can produce cross-linking to strengthen a material as the long molecules combine at various points to form a strong three-dimensional matrix. Surfaces can be modified by exposure to reactive gases, and even heat or stretching can dramatically alter the physical properties of a plastic material. For example, it is common to flame treat a plastic surface to facilitate printing.

Plastics offer some vitally important properties for packaging. They are lightweight, tough, water resistant, inert, hygienic, easily formed into complex or very thin sections, virtually unbreakable, vary from highly transparent to brightly coloured, and can be reprocessed after use or incinerated to allow recovery of much of their energy content.

Although civilization existed without plastics for millennia, there is no doubt that, given the present pattern of living, a vastly greater amount of other resources would be needed if plastics were for any reason not available. The German Plastics Manufacturers Institute (GVM), in response to a high level of criticism against plastics, has produced an assessment of the impact which the abolition of all plastics packaging would have on the German economy. By allocating appropriate alternative materials for each of the plastics materials currently used, it conservatively projected that the total consumption of materials would increase dramatically, by four times as much. Furthermore, energy consumed would double, and the volume of waste would double too.[1]

There is no question that the use of plastics for packaging will grow in the future. Table 5.1 above shows that all plastics were expected to grow by at least 4.5% from 1995 to 2000, with PET expected to experience the highest growth.

Plastics processing

All thermoplastics are melted by heat. Pressure makes them flow and take on new forms; cooling sets the shape. Scrap, defective products and used plastics can all be remelted or recycled. This easy formability and economic recovery makes thermoplastics popular for packaging applications.

After the plastic resin is made into pellets, the first stage of plastics processing is heating the plastic and forming products by either a continuous extrusion process or the intermittent injection moulding process which is described later.

In extrusion, polymer pellets are fed into a hopper by gravity and blended and forced, by the use of a screw extruder, through a heated zone before emerging as a continuous film or tube of molten plastic which can then be cast or blown into sheet or into shapes like bottles. A schematic of a typical screw extrusion process is shown in Figure 5.1. For large packages like drums, the screw extruder is not large enough, and so an accumulator head extruder is fed by a plunger.

For any of the extrusion methods, it is possible to produce multilayer structures, or coextrusions, of two or more polymers, colours and so on. Coextrusion can combine materials with different properties to produce optimal performance. The technique

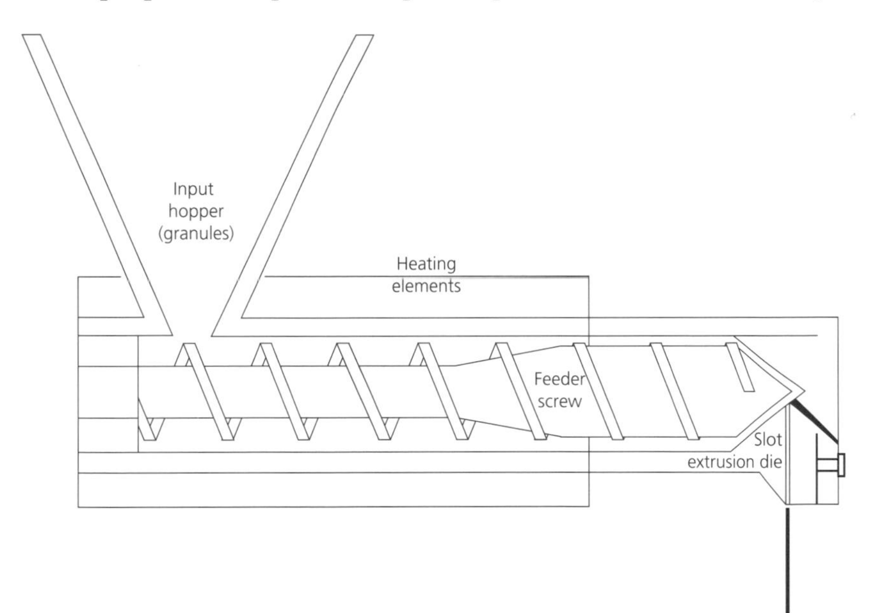

Figure 5.1 Typical extruder

requires only a number of screw extruders (one for each different layer) which can feed their output to a combining die to force the layers out together. The molten layers fuse as a single material if they are compatible.

Film-making

To make sheet film, the plastic is extruded either as a sheet and cast onto chill rolls, or blown into an inflated tube or bubble.

Cast film can be thick or thin; the thicker sheets are generally used for thermoforming or are die cut and scored for boxes. Thinner cast film is used for many packaging applications.

Cast film is cooled more rapidly than blown film, and the film is oriented primarily in the direction of extrusion. Some cast films are biaxially oriented after extrusion, which strengthens them in both the machine and cross-machine directions; at the same time the procedure makes the film thinner. Orientation can also improve clarity, moisture barrier properties and low-temperature durability.

By contrast, blown film is cooled more slowly by air blown to inflate the seamless tube of plastic as it is drawn upwards. This slower cooling allows the molecules to be oriented in all directions, producing a film with greater puncture resistance. Although gauge control is more difficult, uniformity can be improved by using a rotating die. After cooling, the tube is flattened and/or slit and reeled up as a sheet or layflat tubing. A schematic of a typical film blowing process is shown in Figure 5.2.

The ability to heat seal thermoplastic films is one of their most important forming attributes. Table 5.3 shows typical sealing ranges for packaging plastics. A wide heat-seal temperature range is usually desirable to ensure seal quality under variable conditions. Although most plastics will seal given a high enough temperature, high-temperature plastics are generally coated with a plastic which can be activated at a lower temperature, typically LDPE.

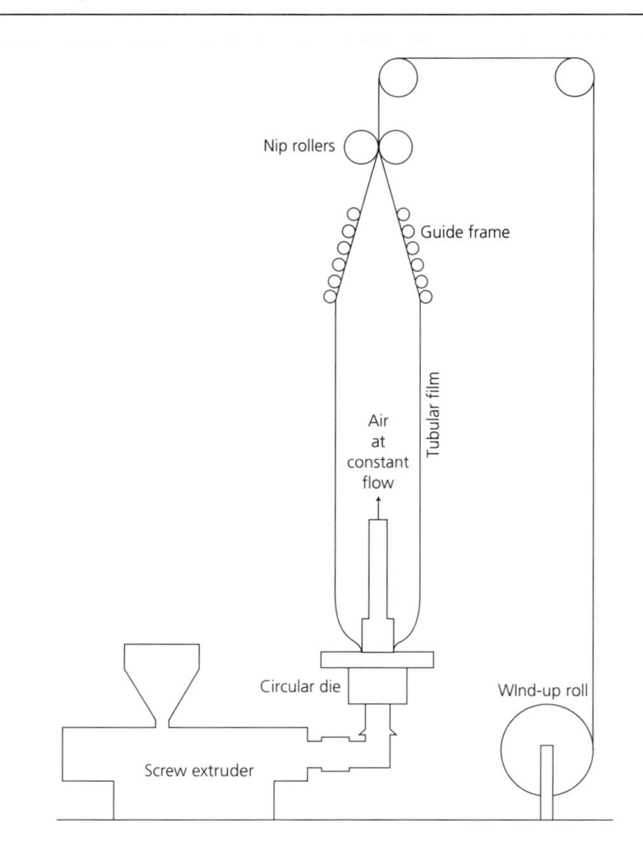

Figure 5.2 Typical film blowing equipment

Table 5.3 Typical temperature ranges at which a plastic can be heat sealed

Material	Heat-seal temperature range (°C)
Ionomer	88–205
LLDPE	121–148
LDPE	121–177
HDPE	135–155
PVC and PVDC	121–177
PP	163–205
Fluorocarbon	177–205
Nylon	177–260
Polycarbonate	205–221

Note: Neither PET nor OPP can be heat sealed without special treatment or coating

Source: Adapted from Soroka, W *Fundamentals of Packaging Technology* IoPP, USA (1995)

Rigid plastic moulding

Bottles are manufactured by a blow moulding process. There are three types: extrusion, injection and injection stretch blow moulding.

Extrusion blow moulding is similar to glass bottle blowing. While still soft, a thick hollow tubular extrusion, called a parison or pre-form, is clamped between two halves of a mould and air is then blown into the neck. The bottle is cooled in the mould and trimmed. Only the outer surface exactly matches the mould and there is a limit to the precision achievable owing to this restriction.

Most plastics can be extrusion blow moulded, including PE, PP and PVC. It is the predominant process for making small to large plastic containers, up to large tanks. Most common detergent and milk bottles are made by this process.

Bottles having two or more layers of material can be made from a coextruded parison. It is common to use recycled material in such a sandwiched construction, with the recycled plastic being flanked by virgin material. Such constructions eliminate possible product contamination from the recycled stock and enable the use of a standard colour for the outer package. Another application is to sandwich a layer of ethylene vinyl alcohol (EVOH), a very good oxygen barrier but which is susceptible to moisture, between two layers of a good moisture barrier. In some cases, a tie or adhesive layer must also be incorporated to improve the material-to-material bond.

The economics of blow moulding (including the transportation of empty containers) are such that many bottles are now formed in, or adjacent to, the facility that fills and labels them. There are increasing efforts to integrate these operations. In new aseptic blow–fill–seal processes, bottles are blown, filled with sterile liquid while still in the mould, and then sealed with a heated die before they are ejected. Other asceptic systems are available in which the inherently sterile bottles are sealed at the point of manufacture with a thin removable diaphragm. This is later cut off under asceptic conditions at the filling point. In-mould labelling is another interesting variation in which a printed label with heat-activated adhesive is placed into the blow mould cavity and bonded to the expanding hot parison.

Injection blow moulding adds precision to the bottle-making process. It begins with an injection moulded tube-shaped parison which is then blown into a second, full-size mould.

Since the neck is injection moulded, dimensions are more precise, an important feature for complex child-resistant and snap-on closures. The injection blow moulding process is used mostly for pharmaceutical and cosmetics bottles because they are small, and precise neck finishes are important.

Injection stretch blow moulding is used to produce the large number of PET bottles used for carbonated beverages and water. An injection-moulded parison is heated to a temperature which just barely allows the parison to be inflated and align its molecular structure. Then the material is heated and stretched by means of a pneumatic rod inserted into the neck before being blown into the mould (see Figure 5.3).

There are two alternative processes available. In the single-stage process, the residual heat of the original moulding process is retained in the preform and it is stretch-blown immediately. In the two-stage process, the preform is allowed to cool and is transported to the point of use where it is reheated and blown into containers at the filling site.

In stretch blow moulding, the plastic is strengthened by biaxial orientation of the molecules, which improves strength, transparency, gloss, stiffness and gas barrier performance. Although most plastics are capable of orientation, PET and PP (and to a lesser extent PVC and PEN) are most commonly processed in this manner.

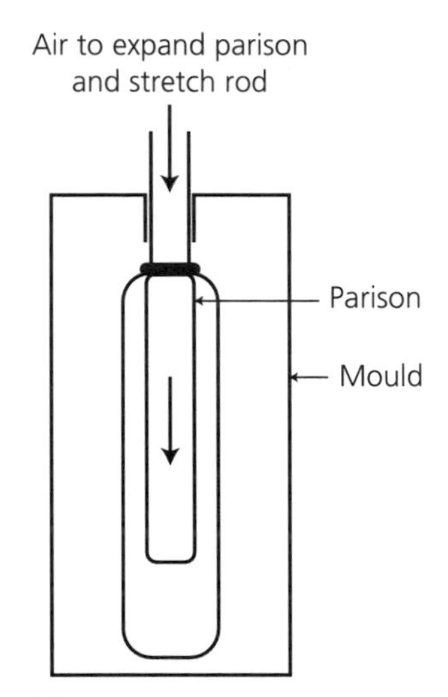

Figure 5.3 Stretch blow moulding

Injection moulding is used for making closures, wide-mouth tubs, boxes, complex shapes, and the parisons for injection blow moulding. The molten plastic is injected under pressure into a fully detailed mould. The mould has two parts which are clamped together during moulding and then opened to eject the item (see Figure 5.4). Highly precise dimensions, very fine detail and very thin sections are all possible using this technique.

Thermoformed containers and parts are moulded from extruded sheet plastics. Thinner (0.5–2.0 mm) materials are made into blisters, cups and trays by a continuous web-fed process. Thicker material (1.5–13.0 mm) is normally sheet fed and used to make pallets and dunnage trays.

The sheet is softened by heat and then forced against a moulded shape or cavity using air pressure (either vacuum or positive compressed air, the former is limited to 1 atmosphere) and/or by mating matching moulds (see Figure 5.5). Then the material is cooled and trimmed. Since material distribution is directly related to the part's geometry, a plug-assist is sometimes used to increase uniformity, especially in the corners, which invariably cause the thinnest sections.

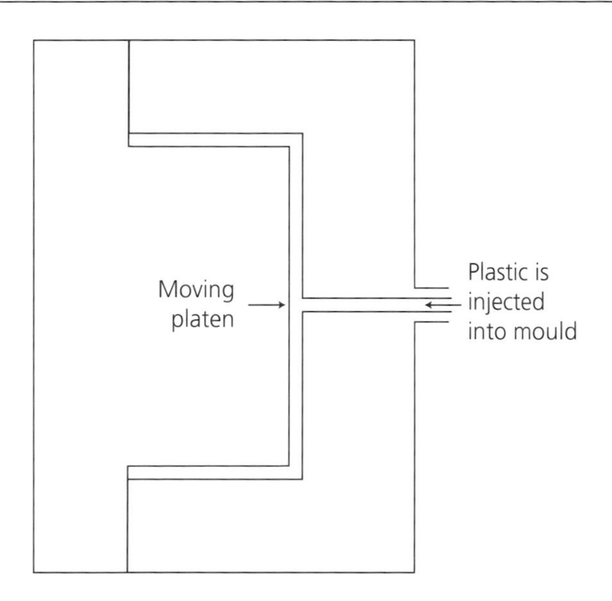

Figure 5.4 Injection moulding

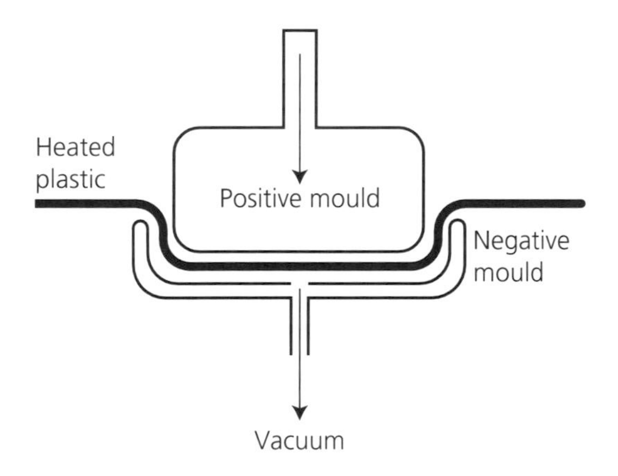

Figure 5.5 Thermoforming

Foamed plastics are formed by dispersing bubbles of gas throughout a fluid polymer and then stabilizing the resulting cellular structure. The material can be extruded in a plank or injection moulded. Foamed plastic applications range from flexible food trays and cushioning materials to rigid structural foam pallet boxes.

6s
Polyolefins — polyethylene and polypropylene

Polyolefins are the workhorses of the plastics packaging sector because of their wide range of useful properties. They are the most inexpensive of all plastics. They are tough, strong and a good barrier to water vapour. As the properties of polyolefins are continually being improved, the value of these materials in packaging applications continues to grow.

As their name implies, polyolefins are formed by the polymerization of certain unsaturated hydrocarbons known as olefins (or alkenes). Polyethylene and polypropylene are by far the most important polyolefins used for packaging, although other members of the family such as polybutylene and polymethylpentene have their own established uses.

Each polyolefin is characterized by its primary building block, with each successive one in the series containing one more CH_2 group. Thus ethylene is C_2H_4, propylene is C_3H_6, and butylene is C_4H_8. The number of hydrogen atoms is always twice the number of carbon atoms. Polyethylene's molecular arrangement is shown in Figure 6.1.

The fact that the atoms of carbon and hydrogen may be arranged in many different ways makes it possible to produce variations in

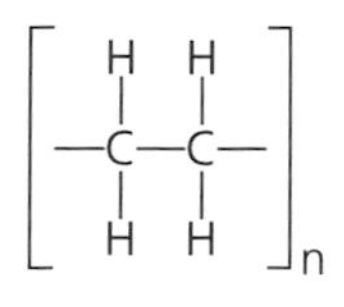

Figure 6.1 Repeat unit of polyethylene

the properties by using different polymerizing techniques and different catalysts. There is currently a great deal of research in the area of new catalysts, especially metallocenes, to improve the properties of polyolefins.

Polyethylene (PE)

Polyethylene is valued for three properties: toughness, heat-sealability and the barrier that it presents to water and water vapour. Other beneficial characteristics are little, if any, moisture absorption and low coefficient of friction. It is generally inert and has excellent chemical resistance, although it is attacked by oxidizing acids and is permeable to gasoline and xylene.

Polyethylene has the lowest cost of the packaging resins. Since it also has the lowest softening point of the packaging plastics, it also has low processing energy costs. The family of polyethylenes is the most versatile and economical of the polymer resins. As a result, polyethylenes are one of the most popular packaging materials; uses range from milk bottles to the ubiquitous plastic bag.

The low softening point, however, makes PE unsuitable for hot-fill applications. PE films and bags maintain their flexibility at low temperatures, and are used for frozen food.

Polyethylene is difficult to print and the surface must be flame treated, for rigid containers, or corona discharge treated in the case of films.

Polyethylene is formed of long chains of C_2H_4 units, but owing to the ability of carbon atoms to form side branches and the variations in conditions in which polymerization takes place, the material is not always formed as molecules of standard shape or the same size. The molecules are long straggling chains with branches, tangled together in various ways to form the tough, transparent, heat-sealable material.

The properties of the several types of commodity polyethylene depend on the density, molecular weight, the morphology (molecule shape) and the degree of crystallinity. The primary performance difference among the types are in rigidity, heat resistance, chemical resistance and ability to sustain loads.

Density is to a large extent a measure of the crystallinity of the material. The density of polyethylene can range from a high of 0.970 g/cm^3 to a low of 0.880 g/cm^3. Table 6.1 shows the relative densities of common polyethylene types, from high to very low density.

Table 6.1 Relative densities of polyethylenes

	g/cm^3
HDPE	0.940–0.970
LDPE	0.915–0.939
LLDPE	0.916–0.940
VLDPE, ULDPE	0.880–0.915

Source: *Modern Plastics Encyclopædia Handbook* McGraw-Hill (1994)

As density increases, so do the properties of tensile strength, gas and water vapour barrier, rigidity and temperature stability. Properties which diminish with increasing density are clarity, impact strength, elongation and heat-sealability.

The molecular structure of the types of polyethylene vary. Low density polyethylene (LDPE) is characterized by long side-branches that give the resins their combination of flexibility, clarity and ease of processing. High density polyethylene (HDPE) has a more linear structure, allowing for a tighter packing of molecules and resulting in a denser, stiffer material. Linear low density polyethylene (LLDPE) lacks the long chain branching of LDPE and has a narrower molecular weight distribution. Figure 6.2 shows how their molecular branching structures differ.

Low density polyethylene (LDPE) was first produced in 1933 in England by Imperial Chemical Industries. In the early 1950s, Phillips Petroleum commercialized the catalysts which are used

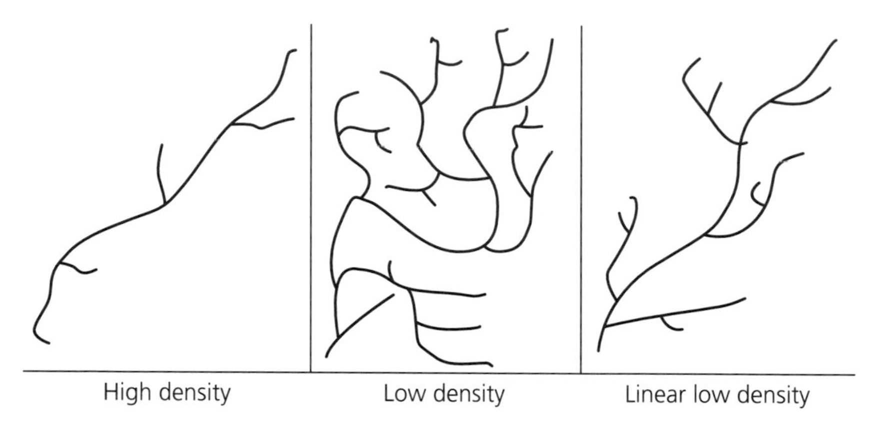

Figure 6.2 High, low and linear low density polyethylene structures

to produce high density polyethylene (HDPE), which became the first commercial product of catalytic ethylene polymerization. In 1960, DuPont Canada began producing linear low density polyethylene (LLDPE) using a new group of transition-metal catalysts. In 1976 a new family of catalysts using metallocene 'single site' catalysts was discovered. Now, HDPE, LLDPE and VLDPE (very low density polyethylene) can be produced by any one of a number of processes to have tailored properties.

High density polyethylene (HDPE)

High density polyethylene is one of the largest volume plastics used in packaging because it is economical and can be formed in a wide range of forming processes. It can be:

- blow moulded into bottles for milk and juice;
- injection blow moulded into cosmetics bottles;
- injection moulded as caps and crates,;
- extruded into film for use in bags of all kinds;
- thermoformed into tubs or pallets;

- rotational moulded into large carboys; and

- foamed for structural foam pallet boxes.

HDPE is stiff, has good tensile strength and heat resistance. Its high density makes it a better water vapour barrier than LDPE, but it is still a poor oxygen barrier. Clarity is poor and the material can usually be recognized by its opaque appearance.

Chemical resistance is good and can be improved by surface treatments, such as fluorination or sulfonation, or by coextruding with higher-barrier plastics like nylon.

It has only moderate environmental stress crack resistance, which can cause bottles of detergent to fracture when stacked. This property can be improved by using a copolymer HDPE with a lower density, reduced crystallinity and higher molecular weight.

Table 6.2 shows a breakdown of the US markets for HDPE. The single largest market is blow-moulded bottles used for milk, cleaners, shampoo, drugs, motor oil and other consumer products. Injection-moulded HDPE is used for cups and tubs.

Most industrial returnable packaging is made from HDPE, including injection-moulded crates, totes, pallets, drums and storage containers. These applications highlight the high strength of HDPE.

HDPE packages can be microwaved, but the material starts to lose its rigidity at temperatures above 93°C, making it suitable only for foods that do not get too hot.

Many retail and food bags are made from extruded HDPE film. It is very strong, a good moisture barrier and is often used to impart moisture resistance. Extrusion coatings of HDPE are used to give moisture and grease resistance to paper packaging materials.

Table 6.2 US packaging markets for HDPE, 1997[a]

Markets	'000 tonnes	%
Blow-moulded packaging		
Liquid food bottles	580	23.5
Household and industrial chemical bottles	478	19.3
Motor oil bottles	80	3.2
Industrial drums	109	4.4
Extrusion films and coating		
Food packaging film	72	2.9
Grocery and retail bags	413	16.7
Extrusion coating	26	1.1
Injection-moulded packaging		
Pails	388	15.7
Tubs and containers (including drinking cups)	132	5.3
Crates and totes	147	5.9
Caps and closures	46	1.9
Total	**2471**	**99.9**

[a]Throughout Part 3, the plastics application tables from *Modern Plastics* magazine report only US consumption because only the US statistics report packaging applications. In the same issue (every January), statistics are given about plastics consumption for other regions of the world, but there is less detail about packaging applications. The percentages shown are the calculations of the authors, based on the plastics applications that we could recognize as packaging and do not include the total consumption of that material for other areas.
Source: *Modern Plastics* (January 1998)

A major advantage of HDPE film is its high melting point, making it suitable for boil-in-the-bag applications. Clarity is generally poor and heat sealing, although achievable, is more difficult than for low density grades. Pigmented grades of thin film are frequently used to make small bags for wet foods like meat and fish.

The difference in the melting points of HDPE and LDPE is exploited in one of the major uses for high density film — the liners for breakfast cereal packs. These consist of a coextruded film of LDPE and HDPE; the heat seal is achieved by the inner surface of low density, but it does not fuse the outer tough high density layer. This allows the two surfaces to be easily peeled apart to open.

In some communities, HDPE bottles are collected for recycling. In the USA, over 250 000 tonnes of recycled HDPE[2] is used annually and the amount is expected to increase steadily.

Recycled HDPE can be made into building materials and new packages. When used to make bottles, the recycled material is often sandwiched between virgin layers in order to overcome the problem of mixed colours, grades and sources. Recycled HDPE is used, with LDPE, to make garbage bags.

Low density polyethylene (LDPE)

Low density polyethylene film is one of the most widely used packaging materials on the market. Its uses range from thin-gauge garment bag film to liners for large water storage tanks.

LDPE is tough, semi-flexible and shock resistant. It is a good barrier to water vapour, but many organic vapours and essential oils will permeate rapidly. It is chemically inert and almost insoluble in all solvents at ambient conditions, but it is susceptible to stress cracking when exposed to surfactants like concentrated detergents.

It is a poor gas barrier to oxygen and carbon dioxide, and where oxidation of a food product is likely, commodity LDPE is not suitable.

Table 6.3 shows a breakdown of the US markets for LDPE. Over 50% of all LDPE is extruded into film, which is then converted into garbage bags, food packaging, retail bags, stretch and shrink film, and industrial bags and liners.

Table 6.3 US packaging markets for conventional LDPE films, 1997[a]

Markets	'000 tonnes	%
Extrusion films and coating		
Food packaging film	460	30.5
Non-food packaging film	382	25.3
Stretch/shrink film	157	10.4
Carry-out bags	49	3.3
Trash bags	29	1.9
Coating	430	28.5
Total	**1507**	**99.9**

[a]See note to Table 6.2 (p 90)
Source: *Modern Plastics* (January 1998)

LDPE (and LLDPE) film can be made by cast or blown extrusion. It can also be extruded as a coating onto another material, and a significant proportion of LDPE (and LLDPE) is used in multi-layer laminated or coextruded structures where the low density material serves as a heat-seal medium. Many laminated structures also use LDPE for clarity and as a water barrier, using other materials (like other plastics or foil) to provide a gas barrier.

Uses, other than film, include injection-moulded containers, extrusion coatings and rotational moulding. LDPE is used for moulded containers where squeezability is needed, such as squeezable condiment containers. Some snap-on caps are also made from LDPE, exploiting its high elongation property. There are now many different grades of LDPE based on the lengths of molecules and their degree of branching and cross-linking.

LDPE can also be foamed to produce cushioning materials which are tough and resistant to creep under load. PE cushioning is more expensive than expanded polystyrene, but its resilience is valuable for reusable applications.

Linear low density polyethylene (LLDPE)

Linear low density polyethylene has a different structure from LDPE, even though the two compete for many of the same flexible film applications. LLDPE has an almost linear molecular structure (hence its name) but it does include short branch chains. The LLDPE resins occupy the middle of the LDPE density range at 0.912–0.928 g/cm^3, but their properties are in most respects superior to ordinary LDPE. A further benefit is that the same reactors can be used to produce LLDPE and HDPE.

Performance benefits of LLDPE are higher physical strength in all respects and higher temperature tolerance. Since there is no long-chain branching, it has much greater elongation than LDPE. LLDPE's higher tear, tensile, and impact strength, along with improved resistance to environmental stress cracking, allow a stronger material to be produced with less material, which has been especially important in film markets. It also offers better strength, durability and chemical resistance than LDPE, but is less transparent.

Metallocene-based LLDPE resins offer excellent clarity and sealability, which makes them especially desirable for applications such as poultry and frozen food packaging.

Typical uses for LLDPE are for carrier bags, stretch films and heavy-duty plastic sacks. It can provide benefits in virtually all polyethylene applications. Table 6.4 shows a breakdown of the US markets for LLDPE.

Table 6.4 US packaging markets for LLDPE films, 1997[a]

Markets	'000 tonnes	%
Extrusion films		
Food packaging film	183	2.5
Non-food packaging film	373	25.6
Stretch/shrink film	328	22.5
Carry-out bags	78	5.3
Trash bags	490	33.5
Coating	9	0.6
Total	**1461**	**100**

[a]See note to Table 6.2 (p 90)
Source: *Modern Plastics* (January 1998)

When introduced, LLDPE was offered at a higher price than LDPE, but since the manufacturing process also offered advantages to the producers, increased capacity became available as many replaced old LDPE capacity with LLDPE manufacturing. This has helped to force prices down. LLDPE was initially blended with LDPE as a cost optimization measure, but as more has become available, its use a single material has grown.

LDPE and LLDPE are often blended with ethylene vinyl acetate (EVA) to improve toughness, heat-seal or cling properties. Most stretch film, used for unitization and over-wrapping, is made from LLDPE, with EVA added to improve stickiness. EVA is discussed further in Chapter 7.

New developments — VLDPE and metallocene catalysts

Plastics manufacturers have continued the development of the polyethylene family as they have learned more about the possibilities for tailoring the molecular structure by using different monomers, processes and catalysts.

Ranges of very low density (VLDPE) and ultra low density polyethylene (ULDPE), having densities lower than regular LDPE, have now become available. Physical properties for these ultra low density grades are superior even to LLDPE, with higher elongation, better puncture resistance and hot tack (making them particularly good for heat sealing through surface contamination), high clarity and enhanced water vapour barrier performance.[3]

VLDPE resins are used primarily in coextrusion or in blends with LDPE, HDPE or LLDPE to take advantage of their properties. Film applications include meat packaging, shrink film and frozen food packaging.

A new generation of polyolefins has been developed, using single-site catalysts called metallocenes. Applications are growing for these tough plastics, which are distinguished by a narrow and highly reproducible distribution of both molecular weight and comonomer composition.

Metallocene catalysts have been found to produce better strength in HDPE, LLDPE, VLDPE and PP as well as PS and other plastics. Moulded containers made from metallocene-based HDPE have twice the impact (drop height) strength of those made from unmodified HDPE. Making LLDPE film with metallocenes improves dart impact, tear and tensile strength. Films have better clarity, lower seal initiation temperature and present a better barrier to moisture and oxygen.[4] Metallocene PE films are used in laminations as a sealing layer for liquid foods.[5]

With such improved properties, some polyethylene materials are beginning to compete in the marketplace with other high performance materials including PVC. Applications include meat, poultry and fish packaging which requires low sealing temperatures, and taste-sensitive packaging such as coextruded structures for cereal and cake liners and coffee pouch laminations.

It is important to remember that, although polyethylene is a common material with relatively predictable properties, all polyethylene is not created equally. Materials with the same specified densities can have very different properties depending on the manufacturing process and catalysts used. As research continues, the properties of this important packaging material (especially its gas barrier performance) have been improved and its applications have been extended. In demanding applications, it is important to be specific regarding the properties needed in order to match the material to the intended use.

Polypropylene (PP)

Polypropylene (PP) is another very versatile polyolefin, used widely for film and moulded containers. Like PE, its polymer structure can be tailored to meet diverse requirements. It has the lowest density of all commercially available polymers, which results in a high yield. It has excellent chemical resistance and good strength at a low cost.

Polypropylene's properties are similar to those of polyethylene, but its melting point at 165°C is higher than that of any of the polyethylene grades, making it difficult to heat seal directly. The high melting point makes PP suitable for microwavable packages (but not for use in conventional ovens).[6]

Polypropylene provides a good moisture barrier, but not a very good gas barrier. Like polyethylene, its properties can be tailored by the selection of catalysts, copolymerization, additives and molecular weight control.

Polypropylene is a semi-crystalline polymer, and the degree of crystallinity (and hence its properties) can be controlled by the manufacturing process. The order and regularity of the monomer units control the end properties of the product. The two ends of the PP molecule are different (see Figure 6.3). If the molecules are linked head-to-head, the polymer has little order and does not crystallize. Such atactic materials are not crystallized, and so they are soft and tacky and are used in hot-melt adhesives. If the monomers are linked head-to-tail, the polymer is called isotactic and is crystalline. Commercial PP usually contains about 95% of the isotactic form which gives PP its density (ranging from 0.895 to 0.920 g/cm^3, typically 0.905 g/cm^3), toughness, solvent resistance, stiffness and heat resistance.

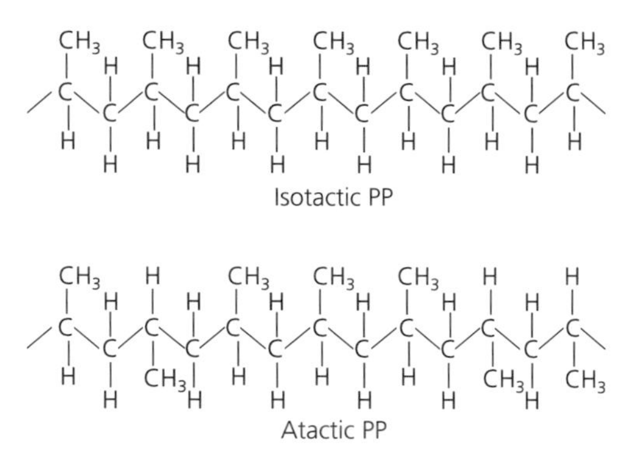

Figure 6.3 Isotactic and atactic polypropylene

Polypropylene is available as a homopolymer or can be blended with another monomer as a copolymer. Random copolymers have small amounts of a comonomer, such as ethylene, at random intervals along the PP chain. These copolymers are relatively clear, have better impact strength and have lower and broader melting points than the PP homopolymer. The lowering of the melting point is proportional to the randomness and the amount of copolymer. Random copolymers are used for blow moulding because of their low temperature toughness. Impact copolymers contain a larger amount of ethylene and are characterized by lower stiffness, enhanced toughness at low temperatures and a relatively opaque appearance.

Substantial modification in properties can be achieved by using additives and fillers. Additives can confer resistance to sunlight, reduce the tendency to retain a static electric charge, and change the coefficient of friction. Fillers such as talc or chalk (calcium carbonate) are used to increase stiffness, improve processing or change the appearance. The use of fillers usually reduces toughness, raises density and increases opacity.

Metallocene-based catalyst technology offers opportunities for further modification. Polypropylene made with metallocene catalysts has better melt flow and lower melt temperatures. The resulting film offers greater yields and thinner films, which can be sealed at lower temperatures. Physical strength properties are also enhanced.

Polypropylene is capable of being converted into the widest range of forms, from monofilament yarns to pallets. Other applications include film (oriented and non-oriented) which can be blown or cast, and containers and closures which can be blow moulded, injection moulded or thermoformed. Table 6.5 shows the major markets for PP in the USA.

Table 6.5 US packaging markets for PP, 1997[a]

Markets	'000 tonnes	%
Blow-moulded containers	77	6.7
Oriented films	429	37.6
Unoriented films	94	8.2
Injection-moulded rigid packaging	541	47.4
Total	**1141**	**99.9**

[a]See note to Table 6.2 (p 90)
Source: *Modern Plastics* (January 1998)

Polypropylene film

Film is the largest application, most of which is in the biaxially oriented form known as OPP (sometimes BOPP). There is some non-oriented PP film used in packaging, primarily for twist-wrap candy.

PP film has progressively taken the market previously held by regenerated cellulose film (cellophane) as the shiny clear overwrapping and form–fill–seal material for snack foods, cigarettes and confectionery. The market for these two alternative materials is often referred to as the Cellopp market.

OPP film is made by extruding a film and stretching it. To orient (stretch) the film, the tubular bubble is inflated or a cast sheet is heated and mechanically stretched by a factor of 300–400% on an apparatus called a tenter frame. Stretching orients the long molecules in both the machine and cross-machine directions, increasing the toughness and strength.

The increase in strength results in a very thin film that is still strong enough for lamination. It gives OPP film the greatest area yield of any packaging film. The film is very strong in tension, but its internal tear strength (ability to propagate a tear that has been started) is very low. This can be a benefit if easy-tear is needed and a starting notch is provided, or it can be a limitation, causing all of the contents of a package to be lost when the film is injured.

Orientation also improves PP's grease barrier properties and low temperature durability. It enhances the material's clarity and gloss. OPP films are stiff; they sparkle and tend to crackle audibly.

OPP's high gloss, high area yield and ability to be made into very thin films makes it the most economical packaging choice for many consumer goods. In addition to cigarette and sweet wrappers, PP film and PP-based laminations are widely used for bags for snack foods and pasta, and as the greaseproof moisture barrier liner in multi-wall paper bags for biscuits and pet foods.

Its high melting point also means that OPP will not heat seal without help. The film can be coated after production with a heat-sealable material such as acrylic, which can simultaneously provide a good flavour and aroma barrier, or PVdC which is also an excellent gas barrier. Alternatively, the film can be coextruded with layers of a lower melting point material. A third option is the use of a hot-melt or cold-seal adhesive.

OPP alone is a good moisture barrier, but a poor barrier to oxygen, light and aroma. Most developments in OPP have been in the area of improving its barrier properties and in exploiting its lamination potential. Other improvements have been to reduce the surface coefficient of friction and its propensity for static generation so that materials can run more smoothly in form–fill–seal machines.

OPP is a stylish material, which can appear in different guises. It is available as an opaque material and can be pearlized or metallized. These forms are growing for special confectionery and snack food applications. Another rapidly growing use is as a substrate for labels, especially foam grades, replacing paper.

Other uses of oriented PP film are for yarns and strapping tape. Yarns, which are made from extruded PP slit into thin ribbons, can be woven into fabric which is used for heavy shipping sacks. Such woven polypropylene is a substitute for jute or burlap, and usually has an ultraviolet inhibitor to protect bags that may be stored outside. PP strapping tape to seal cases or locate pallet loads is being used in competition with straps made from steel, polyester, polyamide and filament tapes.

PP can also be extruded to make corrugated plastic, used in returnable shipping containers. Most corrugated plastic is made from PP, extruded in a profile that resembles a corrugated fibreboard structure, although some is made from PE and some types are made by a similar process to corrugated fibreboard, by laminating two flat sheets to a corrugated sheet.

Moulded polypropylene

Polypropylene is used for blow-moulded bottles and injection-moulded caps, tubs and boxes. PP's reputation as a 'living hinge' has prompted its use for containers and closures where a hinge is an integral part of the design.

Most plastic threaded, dispensing, aerosol and pump closures are injection moulded from PP, as are most thin-walled tubs used for yoghurt and butter. Moulded PP is stiff enough to not deform under the load presented to the thread under torque, but is flexible enough to allow slight undercuts to be moulded to provide a good closure seal. PP caps do not need liners because of the material's resilience. Flip-top caps exploit its living hinge property. Other PP injection-moulded packages include wide-mouth jars, crates, yoghurt cups and cosmetics containers.

PP can be used to blow mould bottles, especially for applications involving aggressive products that stress crack other polyolefins. Clarity of PP extrusion blow-moulded bottles has been a limitation in the past, but newer grades have overcome this problem. The injection–stretch–blow process for the production of bottles and jars, developed initially for PVC and PET, can also be used for producing PP containers. The reheating and stretching of the thick-walled preform has the effect of orienting the molecules in the side walls, providing enhanced toughness and clarity in the same way as these are achieved in PP film.

Where the material's own modest gas barrier properties are adequate, a single layer is suitable; but for bottles which require a higher oxygen barrier, it can be provided by either a central core layer of a high barrier polymer which is co-injected into the preform, or by coextrusion of the parison (in extrusion blow moulding) or by surface coating.

PP has a high melting point and so it can be used for bottles or trays where hot filling or some other thermal exposure method (for sterilizing) is used. Sheets can be thermoformed into heat resistant trays suitable for reheating in a microwave oven. However, PP is brittle at low temperatures, and must be blended with other plastics such as PE when used for frozen food packages.

7
Vinyl-based polymers

The vinyl family of plastics consists of polymers based on either vinyl or vinylidene. It includes polyvinyl chloride (PVC), polyvinylidene chloride (PVdC), polyvinyl alcohol (PVOH), ethylene vinyl acetate (EVA), ethylene vinyl alcohol (EVOH) and polyvinyl acetate (PVA).

PVC differs from polyethylene in having a chlorine atom which replaces one hydrogen atom. Figure 7.1 shows the difference between the molecular repeat units for polyethylene and polyvinyl chloride.

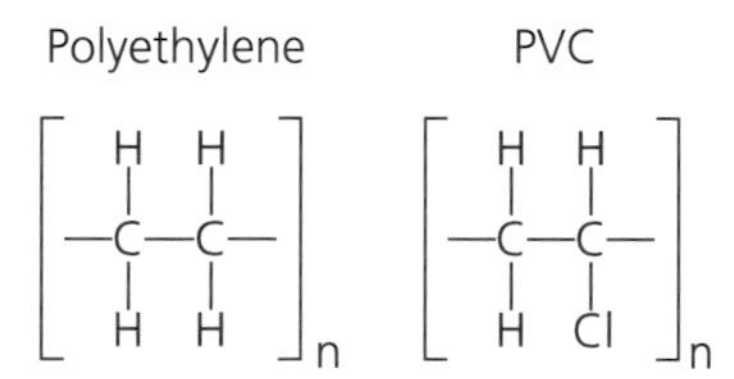

Figure 7.1 Repeat units of polyethylene and PVC

The most important vinyls, from a packaging point of view, are PVC and PVdC. The latter also plays an important role in improving the barrier properties of other plastics.

Polyvinyl chloride (PVC)

Polyvinyl chloride, sometimes called simply vinyl, became popular during the Second World War when it was used as a substitute for scarce natural rubber. PVC is the second largest volume plastic produced. It is used to make many common household products, from shower curtains to credit cards to pipes and building materials like flooring and drainpipes.

Packaging applications account for only about 7% of total PVC sales in the USA. PVC packages include blow-moulded bottles, blister packaging and meat wrapping films. Table 7.1 shows a breakdown of the US markets for PVC.

Table 7.1 US packaging markets for PVC, 1997[a]

Markets	'000 tonnes	%
Film	125	64.8
Bottles	68	35.2
Total	**193**	**100**

[a]See note to Table 6.2 (p 90)
Source: *Modern Plastics* (January 1998)

PVC can be made into a rigid or flexible material. It is tough and clear (has a slight blue tint, and yellows with age), and has good barrier properties. Moreover, it is relatively inexpensive; only the polyolefins and polystyrene are cheaper.

PVC is difficult to process in its pure form, being brittle and unstable. Additives including plasticizers, heat stabilizers, lubricants and impact modifiers are required. Some properties like tensile strength and barrier to gas and moisture depend on the formulation.

PVC is used in two main forms for packaging. One is the rigid unplasticized form (UPVC). It requires the use of antioxidant stabilizers to reduce the thermal degradation which can occur during processing.

UPVC is used for transparent rigid packaging, such as clear sheets for thermoforming, transparent cartons, and extrusion blow-moulded bottles when oil or alcohol resistance are needed. PVC bottles are used for cooking oil, cleaners, chemicals, toiletries and cosmetics.

UPVC is an excellent thermoforming plastic because of its ability to hold a form during the process and its high impact resistance and clarity. One of the most important applications is blister packaging for drugs.

Areas of growth include trays for modified atmosphere packaging, food containers (especially for convenience foods like prepared salads, sandwiches and cooked meats), and transparent cartons. The growing world market for natural mineral waters, most of which are currently packed in PVC bottles, has also been a growth market for UPVC, but there has been an accelerating move from this to PET owing to poor environmental perception of PVC and the falling cost of PET.

Plasticized PVC, on the other hand, is a soft, pliable material which contains a large proportion of a plasticizing compound, usually a phthalate ester.

Plasticized PVC films have excellent stretch and cling properties. They are used widely for hand-wrapping of fresh meats and fresh produce, because the water barrier retards weight loss while the permeability to oxygen allows meat to keep its red colour, and fresh produce to respire slowly.

Less plasticized films are used for a variety of wrapping applications for consumer products like toys. Orientation of film will improve strength, and it is heat-shrinkable. PVC film is also used for medical product packaging, and is suitable for sterilization by irradiation. It is widely used for tamper-evident shrink bands and shrink sleeve labels.

PVC is a good barrier to moisture, gases and odours, but plasticizers reduce these properties. Some plasticizers are toxic and these should not be used with food products. Impact resistance is poor, especially at low temperatures. It must not be overheated as it can degrade and release corrosive hydrochloric acid.

The plasticized grades of PVC have, in years past, been a matter of concern for food safety scientists, since residual vinyl chloride monomer (VCM) has been found to migrate out of the PVC and into wrapped food products. VCM has been found to be a carcinogen under some conditions. In the 1970s, the US government banned the use of PVC for liquor bottles when it discovered a rare cancer in PVC workers. The cause was believed to be the vinyl chloride monomer, traces of which can be released during thermal processing.

However, greatly improved manufacturing techniques have reduced the level of residual VCM in today's PVC to negligible amounts (under 10 ppb). Since the 1980s, PVC manufacturers have changed both the types and the amount of plasticizer and are now also using polymeric plasticizers which have less tendency to migrate.

Although PVC is a widely used plastic, it has not been widely recycled, since most of its uses are for durable goods. The disposal of PVC, especially by incineration, has been a matter of environmental concern.

Both PVC and PVdC (which is discussed in the next section) contain chlorine, which has led to a prolonged debate on their environmental acceptability for incineration. Several European countries have banned their use. There is a fear that the presence of chlorine during incineration of waste can generate hydrogen chloride gas. It can also, during incineration, combine with hydrocarbon volatiles to produce traces of dioxin which increase the impact of acid rain and pose health risks. Dioxin production during paper bleaching has likewise been a concern, as described earlier in Chapter 2.

The criticism is controversial. The PVC defenders point out that in practice only a small proportion of domestic refuse is incinerated, and when it is, there are other sources of chlorine, including food salt, in waste and the chemicals are also naturally present in some vegetable materials like cabbages. Modern incinerators can incorporate flue gas scrubbers which remove most of the dioxin formed. In Japan, there are incinerators that specifically prevent the emission of such compounds. Much of the criticism does not stand up to scientific analysis and has been blamed on the ability of scientists to detect extremely low, 'insignificant' amounts, vastly less than the amounts present in many natural foods.

Nevertheless, the criticism has had an effect on the market's perception of PVC and PVdC, and in many applications they have been replaced by PET, as well as other plastics like oriented polystyrene and polypropylene on the grounds that they are more 'environmentally friendly'.

Polyvinylidene chloride (PVdC)

Polyvinylidene chloride is a copolymer of vinylidene chloride and vinyl chloride. PVdC was developed by Dow Chemical in the 1930s and is often referred to by that company's tradename, Saran. Most PVdC is used for food packaging applications.

PVdC is one of the best gas barrier plastics available. It is also a barrier to moisture and most flavours and aromas, has good chemical resistance, and is heat sealable.

It can be cast or blown into a film. PVdC has been extensively used in the form of hot-water-shrinkable bags for poultry, commercialized by WR Grace as the Cryovac system. Other, lower cost polymers are now increasingly being used for this.

Monolayer PVdC film is highly transparent (with a yellowish tinge). It is soft, strong, and clings to itself. It has long been used for household wrap and food bags, but it is an expensive material for this use compared with competing LLDPE films.

PVdC is used as the barrier component in a number of coextruded sheet materials, from films to thermoformable sheets. The coextrusion process sandwiches a thin PVdC layer, which is sufficient to provide an excellent barrier, between other materials. Using it in this way makes economic sense, as the polymer is very expensive. Another benefit of the sandwich coextrusion is that PVdC does not come into direct contact with the metal surfaces of the extrusion die where it could cause corrosion problems.

Multi-layer films, usually coextrusions with polyolefins, are used to package meat, cheese, and other moisture- or gas-sensitive foods. The ability to withstand the rigours of hot-filling and retorting make PVdC laminations suitable for use in commercially sterilized packages.

PVdC has been used as a coating on all forms of packaging for many years. It can be coated as an aqueous dispersion or applied by organic solvent. Both methods are used, especially for the coating of cellophane and oriented polypropylene (OPP) film. Paper and paperboard can be coated with PVdC where moisture resistance, grease resistance, oxygen barrier and water vapour barrier are required.

PVdC has been applied to the outside of plastic bottles, especially PET and PVC, to increase their gas barrier properties and make them suitable for oxygen-sensitive liquids such as beer, but these have had technical problems and have not been commercially successful. Such coating is usually achieved by a dipping technique although spray coating and roller application techniques have also been demonstrated.

Although PVdC was the earliest specialist barrier polymer and now has a number of competitors, it was holding its position very well up to the mid-1980s, especially since (unlike some of its competitors) it provides an excellent barrier to both water vapour and oxygen.

PVdC has been subject to some of the same criticism as PVC. Since the 1980s, the continued pressure on environmental grounds has led to it losing market share to alternatives such as EVOH and acrylics.

Polyvinyl alcohol (PVOH), ethylene vinyl alcohol (EVOH) and ethylene vinyl acetate (EVA)

Polyvinyl alcohol is the most commonly used water-soluble film. It is used to package dry products such as detergent and agricultural chemicals, which are added to water in the package. It takes about one minute to dissolve in water. When used with detergent powders, the film enhances the detergent by suspending the dirt in solution.

PVOH is a good gas barrier and resists most chemicals. It is stable at moderate but not high humidity conditions. It is heat sealable.

Applications include disposable bags used in hospital laundries to reduce the possibility of cross-infection. Other special applications in packaging include unit dose packages for difficult or hazardous materials such as powder dyes or agrochemicals.

Ethylene vinyl alcohol (EVOH) is a copolymer of ethylene and vinyl alcohol. It was developed to overcome the moisture sensitivity of PVOH. The moisture sensitivity depends on the proportions of ethylene and vinyl alcohol in the copolymer, the higher the percentage of ethylene, the better the water resistance but the worse the barrier. In EVOH, a compromise level is chosen.

EVOH is best known for being an outstanding barrier to gases such as oxygen, carbon dioxide and nitrogen. EVOH is therefore the right choice for many food packaging applications, capable of maintaining a modified atmosphere as well as preventing oxidation. It is highly resistant to hydrocarbons and organic solvents. This makes it a good choice for packaging oily foods, edible oils, pesticides and organic solvents. It is also a good aroma barrier.

EVOH resins were first commercialized in Japan in the early 1970s, and became more widely used in the mid-1980s when US food producers started using the resin for all-plastic, squeezable bottles.

The primary use for EVOH is food packaging. In most cases, it serves as an oxygen barrier ply in a coextruded or laminated film. Flexible films are the largest volume application for EVOH, used for packing processed and fresh meats, coffee, condiments and snacks. Depending on the structure, EVOH-based packages can

Table 7.2 EVOH structures and applications

Fabrication process	Application	Structure[a]
Cast coextrusion	Processed meats, natural cheese, snacks, bakery	PP / nylon / EVOH / nylon / LLDPE Nylon / nylon / EVOH / nylon / Surlyn Nylon / EVOH / Surlyn PET / LDPE / EVOH / Surlyn
Blown coextrusion	Processed meats, bag-in-box, red meat, pouches	Nylon / LLDPE / EVOH / LLDPE LLDPE / EVOH / LLDPE LLDPE / EVOH / LLDPE / Surlyn Nylon / EVOH / LLDPE
Lamination	Coffee, condiments, snacks, lidstock	OPET / EVOH / LDPE / LLDPE OPET / EVOH / OPET / PP
Coextrusion coating	Juice, bakery, laundry products	LDPE / paperboard / LDPE / EVOH / LDPE / LDPE / paperboard / LDPE / EVOH / LDPE / EVOH OPP / LDPE / EVOH / LDPE / EVA
Thermoforming	Vegetables, fruit sauce, entrees, pudding	PP / regrind / EVOH / regrind / PP PS / EVOH / LDPE
Coextrusion blow moulding	Ketchup, sauces, cooking oil, salad dressing, juice, agricultural chemicals	PP / regrind / EVOH / PP HDPE / regrind / EVOH / HDPE HDPE / regrind / EVOH PET / EVOH / PET / EVOH / PET
Profile coextrusion	Cosmetics, toothpaste condiments, pharmaceuticals	LDPE / EVOH / LDPE LDPE-LLDPE / EVOH / LDPE-LLDPE

[a]Tie layers are omitted for clarity

Source: Foster, R 'Ethylene-vinyl alcohol copolymers (EVOH)' *The Wiley Encyclopædia of Packaging Technology* (1997) p 359

even be hot-filled and retorted, although there may be a reduction in oxygen barrier performance. EVOH is coextrusion blow-moulded into bottles for ketchup, sauces, salad dressing and juice. It is also used with LDPE to produce collapsible tubes for toothpaste, cosmetics and pharmaceuticals. Coextruded sheets can be thermoformed. Table 7.2 shows some typical multi-layer EVOH structures and their applications.

Since EVOH is still to some degree moisture sensitive, it is usually flanked by polyolefins or another good water vapour barrier such as nylon. Heavyweight films include outer layers of polystyrene, polyvinyl chloride or polyester. Tie layers are needed for all plastics except nylon. To further improve moisture resistance needed for packaging retorted food, a desiccant (to absorb moisture) can be incorporated in the tie layer.

EVOH is also a popular high-barrier coating, shielding from gases, oils, odours and organic solvents. It is applied by spraying, dipping or roller techniques. It is used as a coating on paperboard — to replace a foil barrier — for juice, bakery and laundry products.

Ethylene vinyl acetate (EVA) copolymers of vinyl acetate and ethylene have similar properties to LDPE. EVA is often blended with LDPE to improve stretch, heat-sealability and cling. It is used more extensively in the USA than in Europe. EVA's properties depend on the proportions of vinyl acetate to ethylene in the final blend. These normally vary from about 5% to 50%. At vinyl levels of about 20%, the material is like plasticized PVC, soft and clingy, and is used for stretch film or as a heat-seal layer in coextrusions. An 8% vinyl content produces a material like LDPE, but with better toughness, elasticity and heat-seal strength. When present as a small constituent in polyethylene (below 5%), the vinyl acetate is primarily an additive to help improve processing and heat-sealing performance.

EVA is coextruded with other materials to improve strength, stress-crack resistance and heat-sealability. It can be extrusion laminated to metallized polyester for bag-in-box constructions for liquid. Primal meat cuts are vacuum packed in a coextrusion of EVA and PVdC.

EVA copolymers are a primary component of hot-melt adhesives. Polyvinyl acetate (PVA) is also a commonly used vinyl-based adhesive for paper. These adhesives are further discussed in Chapter 14.

8
Styrenic plastics

Polystyrene is the most common styrenic plastic used for packaging, but there are a number of other copolymer plastics based on styrene, including acrylonitrile butadiene styrene (ABS), styrene acrylonitrile (SAN) and styrene butadiene (SB). Styrene can be copolymerized with other monomers in order to obtain a wide range of properties.

The styrene monomer is quite different from the olefin type, as it is based on an aromatic molecule (one containing the benzene ring in its structure). It forms in chains that resist rotation, yielding a stiff, brittle material. Styrenes are unable to crystallize and so are highly transparent.

Polystyrene (PS)

Polystyrene resin is one of the most versatile, easily fabricated and cost effective plastics. It can be moulded, extruded and foamed. It is widely used to make sturdy but disposable dishware, jewel boxes, food trays, closures and cushioning. Table 8.1 shows a breakdown of the US markets for PS.

In some communities where economics or legislation is favourable, PS packaging, especially foamed PS, is recycled. When incinerated, PS, like PVC, causes unacceptable gaseous emissions, and special scrubber equipment is required.

Two types of polystyrene are available: general purpose and high impact.

Table 8.1 US packaging markets for PS, 1997[a]

Markets	'000 tonnes	%
Moulding (solid PS only)		
Closures	56	5.6
Rigid packaging	54	5.4
Produce baskets	16	1.6
Extrusion (solid PS only)		
Oriented film and sheet	167	16.6
Dairy containers	89	8.8
Vending and portion cups	152	15.1
Lids	80	7.9
Extrusion (foam PS)		
Food trays	110	10.9
Egg cartons	30	3.0
Single-service hinged container — SS	55	5.5
Expandable bead (EPS)		
Packaging shapes	62	6.2
Cups and containers	91	9.0
Loose fill	44	4.4
Total	**1006**	**100.1**

[a]See note to Table 6.2 (p 90)
Source: *Modern Plastics* (January 1998)

General purpose polystyrene

General purpose polystyrene is a glossy, highly transparent, non-crystalline polymer (despite its frequently used name, crystal polystyrene, which refers to its clarity and hardness rather than its structure). Its surface is smooth and shiny. It has a density of 1.05 g/cm^3 and softens at about 95°C.

General purpose PS is brittle, which has restricted its use in packaging mainly to thick, clear, injection-moulded containers like 'jewel' boxes used for hardware, audio tapes and CDs, toys, cosmetics, and (as the name implies) jewellery.

It is a poor water vapour and gas barrier with low heat-seal strength. Blow-moulded bottles are used for talcum powder. Non-packaging applications include disposable medical devices, cutlery and drinking cups.

Most PS research has centred on improving the physical performance to reduce its brittleness. Metallocene technology could in the future play a role in improving PS by increasing its strength and decreasing brittleness.

General purpose PS can be extruded as a clear film, but this also tends to be brittle (it has a characteristic metallic sound when rustled) and has found only small applications in packaging. An example is the wrapping of flowers and certain fresh produce such as lettuce where the film's high permeability (breathability) helps to restrict wilt by controlling moisture loss.

Two kinds of PS foam are made from the general purpose grade. Foaming reduces the brittleness of PS and capitalizes on its rigidity. PS foam is the most widely used packaging foam, used for cushioning, insulating and void filling. In recent years, it has received increased competition for these applications from PE bubble film, moulded pulp and void-filling air sacs.

Blowing agents, which expand when heated, are easy to incorporate into PS. When heated, the expanding gas gives a cellular structure to the PS. The original fluorocarbon blowing agents, which were controversially linked to the depletion of the earth's ozone layer, have been replaced with hydrocarbons.

Extruded polystyrene foam is made in a sheet which can then be easily thermoformed. This material has good cushioning and insulating properties, and is used for meat/vegetable trays, egg cartons, fast-food containers and as a protective label material for glass bottles.

Expanded polystyrene (EPS) foam is moulded from pre-expanded beads. EPS is one of the most common cushioning materials, used to protect fragile products like appliances and electronics. It is moulded into insulated boxes for fresh fish. It is also moulded into small shapes to be used as a dunnage material or locating fitment, including 'loose fill', to fill voids in packages and add protection to edges and corners.

Biaxially oriented polystyrene film is also less brittle than general purpose PS. It is a clear, sparkling sheet that can be thermoformed into clear, tough items like blister packaging and trays for confectionery, salads, biscuits and condiments. It competes with PVC and PET, and although it is more expensive, it has a higher yield because of its lower density.

Oriented PS has a narrow thermoform temperature range (110–125°C), narrower than PVC can tolerate. High mechanical pressures are used in the thermoforming operation to minimize its tendency to shrink as it nears its melting point. Heat resistance of PS is a restricting factor, and efforts to make OPS more heat resistant continue, aiming for the microwavable meals market.

High impact polystyrene (HIPS)

High impact polystyrene has a small amount of rubber-like polybutadiene or styrene butadiene blended in to overcome the brittleness problem of general purpose PS. The material is tougher, but is less clear, usually translucent or opaque.

Thermoformed HIPS is used in the packaging of foods but must be processed with care to avoid taint problems. Some applications include cups and tubs for refrigerated dairy products, single-serving cups, lids, plates and bowls. It is also used in multilayer extrusions which can be thermoformed to make containers for aseptic food packaging.

A recent development is the addition of polyphenylene oxide (PPO) to HIPS, which has been found to improve heat resistance, toughness and strength. PS/PPO packages are microwavable.

Styrene copolymers — ABS, SAN and SB

Tough copolymers of styrene are available for packaging applications. One is ABS (acrylonitrile butadiene styrene), a tough thermoformable material. It is a copolymer of styrene and acrylonitrile with the butadiene finely dispersed and trapped within the molecular matrix.

By varying the proportions of the three components, a wide range of properties can be obtained. ABS polymers can have good chemical resistance, are tough and hard, resistant to scuffing and staining, and have very good impact, tensile and flex strength. ABS can be either translucent or opaque; the base resin has a yellowish colour.

ABS is easily thermoformed and moulded. Major use areas are for consumer durables like refrigerator door panels and automobile parts. Its high impact strength makes it useful for tote boxes and trays, especially large ones, because it has a low tendency to warp.

There are grades which are used for packaging, mainly as thin thermoformed margarine tubs or trays and cosmetics packaging. However, compared with competing packaging materials, the cost of ABS is high for such consumer packaging applications.

Styrene acrylonitrile (SAN) copolymer is another material which can have packaging applications, the largest of which is as a component in the manufacture of ABS resins.

SAN is clear, rigid and glossy, and it is offered as an alternative to ABS, PVC and OPS for cosmetics packaging — bottles, overcaps, closures and spray nozzles — where clarity is an important feature. Its characteristics are determined by the ratio of styrene to acrylonitrile (frequently 3:1). It is not a particularly good gas barrier, but this can be improved by increasing the acrylonitrile constituent.

Styrene butadiene copolymer (SBC) is a reasonably tough, transparent material with a low density. It is more expensive than polyolefins, but less costly than polystyrene, when competing for similar applications. Drawbacks include a relatively high permeability to moisture and gases, and a tendency to stress crack in the presence of fats and oils.

SBC is frequently blended with other compatible resins like PS and PP to enhance their performance, contributing stiffness, hardness, toughness, strength and high optical properties. PS/SBC blends are used in single-service food packaging, bottles, blister packs, overcaps and film. SBC is also blended with PS to form HIPS (see above).

SBC can be converted by all processing routes into containers, sheet, film and so on. It can be made into bottles, film and thermoformed containers for food and medical products. It is widely used in medical packaging because it can be sterilized by both gamma irradiation and ethylene oxide.

SBC blown film is highly permeable and is used for wrapping fresh vegetables. Injection-moulded containers can have a flexible hinge, similar to those made from PP. It is better known by its tradename K resin and is more widely used in the USA than in Europe.

9
Polyesters

Polyesters are the fastest growing group of plastics used in packaging, primarily because of their widespread use in large bottles for carbonated soft drinks.

The term polyester covers a wide range of materials. The earliest use was as a textile fibre; clothing, carpets and soft drinks bottles are all made from PET. Boat hulls and fishing rods are made from glass-reinforced thermoset polyester commonly called fibreglass. Polyesters are the product of a reaction between an organic acid and an organic base, and it is the thermoplastic types which are of interest for packaging

Polyethylene terephthalate (PET)

The primary packaging use for polyester is as polyethylene terephthalate. PET is inert and has high clarity. It is strong, tough and, in the moulded form, stiff.

It is a relatively good gas barrier and tolerates fairly high temperatures. These properties can be improved by orienting, coating or copolymerizing. It is one of the more expensive plastics and is used when its superior properties are needed. There is no restriction on its use for food contact, and most of its applications are for food.

It is used for rigid containers like bottles, trays, blisters and jars, as well as high-performance films. The primary applications are shown in Table 9.1.

Table 9.1 US packaging markets for PET, 1997[a]

Markets	'000 tonnes	%
Soft drinks bottles	830	54.6
Custom moulded bottles	600	39.4
Clear thermoformed sheet	59	3.9
CPET trays	25	1.6
Coated board	8	0.5
Total	**1522**	**100.0**

[a]See note to Table 6.2 (p 90)
Source: *Modern Plastics* (January 1998)

PET bottles

The highest usage is in bottles used for soft drinks and water. PET has replaced PVC in a number of applications for environmental reasons and where clarity is a prime concern. The use of PET rigid bottles expanded in the 1970s as the result of a strategy by Coca-Cola to increase the sales of soft drinks. Larger containers were desired, but problems with the weight and safety of large glass bottles stimulated the search for alternative materials. A filled PET two-litre bottle weighs 24% less than a similar glass bottle.

At that time, the excellent physical performance of polyesters was well known and the barrier properties, although only modest, were felt to be acceptable for drinks with a short shelf-life. Performance was further improved by the newly introduced biaxial orientation effect of stretch–blow technology for making plastic bottles, which had initially been demonstrated with PVC. Moreover, the availability of the material was increasing owing to a decrease in its demand in textile applications.

Since bottles for carbonated soft drinks have to withstand high internal pressures (up to 4 atmospheres or 60 lb/in^2), flat bases are not possible because they would blow outwards. Therefore, the first generation of PET soft drinks bottles had hemispherical bases which were made stable by the addition of a separate base cap. Later developments using a 'multiple dome' design withstand the internal pressure while still providing a reasonably stable base.

Loss of carbon dioxide through the bottle walls does occur but the rate has been found to be acceptable to retailers and manufacturers. Trials to reduce this by coating the outside surface with PVdC copolymer barrier resin met with mixed results; small traces of carbon dioxide permeating through the PET became blocked by the barrier layer and concentrated in the form of small bubbles or blisters on the surface. Although better coating techniques can eliminate this problem, it was determined that there was no need for such a coating, since it is now widely accepted that a soft drink's carbonation level can fall by up to 15% over a 90 day period, well within the shelf-life capability of PET.

PET was first used for large sizes (1–2 litres) since this range was most economical and performance-effective. Smaller bottles took longer to be adopted; since the carbon dioxide barrier is a function of surface area, small bottles have a higher rate of carbon dioxide loss and coatings are required. PET is now accepted on both economic and performance grounds for bottles as small as 250 ml.

Wine, spa water (plain and carbonated) and toiletries are now sold in PET bottles. One aspect which initially limited the use of PET for the ultra taint-sensitive pure mineral waters was the inevitable presence of minute traces of acetaldehyde in bottle walls. The organoleptic (taste) effects of these minute traces are heightened by the presence of carbon dioxide. This problem has now been virtually eliminated by improved manufacturing techniques which reduce the quantity of acetaldehyde to tiny amounts. Any such residual traces are driven off in the heated stretch–blow processing stage. Now, even carbonated spa waters are satisfactorily packaged in PET.

PET bottles are used for alcoholic beverages, from 50 ml airline liquor mini-bottles to 30 litre party packs. Bottles for edible oil are increasingly being made from PET rather than PVC. It resists weak acids, bases and most solvents.

PET is also used in some countries for beer bottles, but here the critical need is to prevent the ingress of oxygen rather than just the loss of carbon dioxide. Oxygen causes beer to go stale. In order to reduce this to an acceptable degree, PVdC barrier resins have been applied as an external coating. Alternatively, a coextrusion technique is used to incorporate a barrier component such as EVOH or MXD-6 amorphous nylon[7]. The latter approach is the more successful, but is more expensive. Since the mid-1990s, the use of PET/PEN blends, or even all PEN (polyethylene naphthalene dicarboxylate, has been proposed for beer bottles, since these can be designed to be returnable and thus justify their high cost.

PET wide-mouth jars are used for foods, especially for dried products and those which are not hot filled. Some very high quality thick-walled bottles and jars made from PET are used in Japan for cosmetics and toiletries. Many different effects — such as colours, prismatic cut glass faceting, pearlescent and frosted finishes — are achievable. These luxury containers are very expensive and have not been adopted on the same scale outside Japan.

A normal PET stretch-moulded bottle cannot be filled with a hot product. Above 60°C it will distort or shrink since the second processing stage leaves the container with a heat retraction memory. The most critical area is the neck since it must fit the intended closure accurately.

Various methods have been used to improve heat stability. Most involve increasing the material's crystallinity with heat. A dual-mould method heats and shrinks a moulded bottle, and then blows it in a second mould. Alternatively, the bottle can be held in the mould for long enough to relieve the stresses. There are also methods to selectively stabilize the critical neck area. One is to subject the neck zone to a further heat treatment which crystallizes the polymer (CPET). Alternatively, a second, high melting point resin (like polycarbonate or polyacrylate) is used in the neck, or the neck thickness can be increased. Grades of PET with higher temperature tolerance are being developed, and hot-fill temperatures to 85°C and above are now possible.

The profile of the bottle can also be modified to accommodate higher temperatures, as demonstrated in the unit portion packaging of jams and preserves which are hot filled using standard PET grades. Despite the product temperature being higher than should be tolerable, the mass of the small cold jar as a proportion of the weight of jam filled cools the product sufficiently during filling.

Besides heat resistance problems, hot-filled PET bottles are susceptible to vacuum collapse when the product cools. To reduce the effect of the distortion, hot-filled PET bottles are moulded with vacuum panels designed to distribute the distortion uniformly around the bottle.

Bottles made from PET are the most highly recycled of all plastic containers. Over 30% of PET bottles in the USA are recycled, primarily owing to deposit systems in 'bottle bill' states. Reclaimed PET is in great demand for applications such as fibrefill, textiles, carpet, thermoforms, non-food containers and strapping. PET can also be depolymerized (methanolysis) to revert the polymer to the original monomers, which can then be repolymerized. Market testing of a PET 'can' with a metal top failed in the USA and many other countries in large part owing to environmental pressures that favour single material packages for recycling. This pack remains in limited use in Europe.

PET films

Biaxially oriented PET is a thin, high performance film. It was originally developed for recording tapes and is used mostly in low gauges, typically 12 μm.

PET film has exceptional tensile strength, is dimensionally stable, clear and stiff. It an good aroma barrier, but its moisture and oxygen barrier properties are only moderate.

It is not heat sealable, but it can be coated or solvent-sealed. It has excellent thermal properties, making it a good choice for food which is filled hot or cooked in the package. It tolerates temperatures ranging from –70 to 150°C for several hours, and can withstand even higher temperatures for a short period of time.

PET film is often coated with PE, which provides sealability, or PVdC, which improves moisture barrier properties. Such materials are used in boil-in-bag applications, packaging processed meats such as sausages, and as a lidding film for sealed trays which can be heated in a microwave or conventional oven. Thicker, reverse-printed PET film is commonly used in the outer layer of multi-layer stand-up pouches to provide thermal stability during sealing.

As well as providing high strength for a number of multi-layer structures, PET film is an ideal substrate for the vacuum metallization process. When coated with a minutely thin layer (one-millionth of an inch) of aluminium, vaporized under high vacuum conditions, the oxygen barrier improves by a factor of 100–1000, depending on the quality of the metallizing and the thickness of metal deposited. Metallizing also produces a decorative shiny surface.

Metallized PET is used for coffee brick packs, bag-in-boxes for liquid products and bags for snack foods like potato crisps which require a thin but excellent oxygen barrier. Such snack foods have extremely high surface area and high fat content. They are therefore particularly susceptible to oxidation rancidity. This mechanism is accelerated by light, so the opacity provided by metallization is a second important benefit. The most common bag-in-box material is a three-ply laminate of EVA/metallized PET/EVA. Another use for vacuum metallized PET is for susceptor film used in microwavable packages to impart crispness to food. The metallization process is discussed in more detail in Chapter 13.

PET film is easy to print. It is used as a label material, including metallized labels. It is also used as a hologram substrate, initially chosen as a material for credit cards with holographic security labels.

PET film is also used in high performance retort pouch structures in combination with aluminium foil and HDPE or PP. Retort pouches are like flexible cans, in which food is cooked after packing. This material, which can withstand both the thermal and physical stresses of pressure retorting, is used extensively in Japan and some parts of Europe. Retort pouches have never been very popular in the USA and the UK despite high expectations during the 1970s. A similar structure is, however, used for some medical packaging applications.

PET film is relatively expensive, but there are savings in processing through form–fill–seal machines because they can be run at higher speeds without the distortion which occurs with other materials.

Thermoformed PET — APET and CPET

Amorphous PET (APET) can be extruded into sheets and used for thermoforming applications. The resulting material is calendered for high gloss and is more expensive than other thermoformable plastics. Medical devices are often packaged in APET thermoforms.

The same grades of APET are finding applications for the production of transparent cartons. Sheet material is creased, die cut and side-seamed by adhesion or heat. The packs offer an attractive appearance for toiletries, textiles and small household items.

In most such applications, APET competes directly with PVC — from which it is visually indistinguishable — and with sheet PP — which although more cloudy is making some inroads into the transparent carton market. Although PET is more expensive than PVC or PP, it has a faster cycle time in thermoforming, it is easier to use in-plant scrap since thermal degradation is less of a problem, and no stabilizing additives are needed.

Fully crystallized PET (CPET) is less subject to deformation under stress, especially at high temperatures, but is brittle at cold temperatures. It used to make thermoformed dual-ovenable trays. This is a high growth market, related to the growth in microwave oven ownership, which is actively pursued by retailers and food manufacturers who provide ready-prepared (frozen or refrigerated) meals. In some cases, dual-ovenable trays have an APET/CPET structure, with the CPET providing rigidity and the APET providing low-temperature impact strength.

CPET is manufactured using a traditional thermoforming process for thick sheet but the tray remains in the forming mould for a few seconds to achieve crystallization. The effect is visible in that the transparent tray turns opaque white. CPET sheets can also be extruded in a foamed form, resulting in a lighter-weight tray, which has been used for baked goods. CPET trays can tolerate temperatures up to about 220°C.

Paradoxically, if only microwave oven tolerance were needed, PET would not be considered since thermoformed PP trays are less expensive and are quite capable of withstanding the 100–110°C temperatures generated by microwave ovens. However, to provide customers with maximum convenience and flexibility, food manufacturers recognize that a tray suitable for both microwave and conventional ovens is to be preferred.

High performance polyesters — PCTA, PETG and PEN

The complex chemistry involved in the synthesis of the polyester group of plastics allows for the possibility of many further variations offering particularly attractive benefits.

Even higher temperature grades of PET are now available. PCTA copolyester (cyclohexanedimethanol and terephthalic acid copolymer modified with another acid) is a crystallized material with a very high melting point, used for dual-oven trays. It can be alloyed with other plastics or filled with glass fibres or mica to meet a variety of performance criteria.

PETG copolyester is a glycol-modified polyester. In sheet form, its melt temperatures range from 230 to 250°C. It can be extrusion blow moulded into clear bottles, extruded into film and sheet for thermoforming, or injection moulded. Both PETG and PCTA are sterilizable by both ethylene oxide and gamma rays.

PETG is injection moulded into thick-walled jars for cosmetics, both clear and translucent, which look and feel like glass and have excellent resistance to oil and aromas. It can also be used to make packages for food and detergent, as well as trays for medical devices. The material is of high gloss and clarity and has very good processing characteristics. It is easy to print with high quality graphics, including metallic foil transfer. It competes directly with PVC, and although PETG has a superior visual appearance, it is significantly more expensive. Improvements in standard PET and PEN make this material of diminishing interest.

PEN (polyethylene naphthalene dicarboxylate) has superior barrier properties, ultraviolet resistance and temperature stability, making it particularly suitable for hot-fill food applications. Rigid PEN bottles are suitable for use in returnable/refillable applications because they are heat resistant and can be resterilized for re-use.

Compared with PET, PEN provides approximately five times the barrier for carbon dioxide, oxygen and water vapour. It is stronger than PET and UV resistant. Its high-temperature performance is better, which enables products to be hot filled without side-wall distortion, which is a problem for PET.

PEN can be moulded and PEN bottles can be used for foods as well as alcoholic and carbonated beverages. PEN bottles are beginning to be used for beer, a demanding application which requires a good oxygen and ultraviolet barrier.

PEN can be blended with PET, and PET/PEN copolymers can be produced. Such combinations optimize the high cost PEN material and are expected to grow in food packaging applications.

PEN films are stiffer than PET, simultaneously providing a thinner material with a better barrier. Applications include stand-up pouches, modified atmosphere and home meal replacement packaging.

Pure PEN can be easily separated from other plastics because it fluoresces, which will facilitate recycling if it becomes widely used. Recycling is an important option for such a high cost material.

The first PEN applications were in Japan and South America, but there is active interest worldwide for this high performance plastic. Because of its current high cost, it has been used only for food packages where high barrier or high temperature use is essential. However, the price is falling owing to the use of a new, less expensive base material and increased production capacity. PEN is expected to approach cost competitiveness with lower performance plastics.

The potential for new classes of polyester, as exemplified by the PEN material, is such that they could offer the possibility of being the 'ultimate plastics'. They are inert, require no separate additives, are fully recyclable and can either match the performance of traditional materials or at least provide an adequate level of performance for modern food distribution.

10
Nylon (polyamide)

The group of polyamides, or nylon (formerly a DuPont tradename, although it was developed in the UK), comprises a class of chemicals developed in the 1940s. Like polyesters, they were used initially for textiles. A few of these have found packaging applications, although they are usually for highly specific applications where their high gas barrier and strength properties merit the expense.

Nylon type 6 and subtype 6.6 are the most important for packaging. Polyamide chemistry is complex, and a system has been developed for naming the types based on the number of carbon atoms in the original monomer, which represents the size of the repeat group in the long chain polymer. Nylon-6 has six carbon atoms, and nylon-6.6 is a subtype of nylon-6.

The properties of nylon which are of greatest significance for packaging applications are its toughness and strength over a broad temperature range, puncture resistance, grease resistance, resistance to stress cracking, and barrier to gases, oils, fats and aromas. Nylon absorbs water and has relatively poor water vapour transmission properties, but this can be improved by applying a PVdC coating.

Nylon can be cast or blown into film, blow moulded or thermoformed. Nylon is expensive and is often used in coextruded structures with other plastics. Blow-moulded bottles for hard-to-hold chemicals, like toiletries and household cleaners, are made by using nylon as an outer layer in extrusion blow moulded plastic bottles, which provides a highly attractive glossy surface. This may also be pigmented to give a coloured layer over a more inexpensive commodity plastic such as HDPE.

Nylon film

Much of the nylon in packaging is used in the form of multi-layer films produced either by adhesive lamination, when the oriented form may be used, or as coextrusions, usually with polyethylene or polypropylene. When coextruded with polyolefins, as long as the melt viscosities are accurately matched, the materials do not require an intermediate tie layer. Very thin oriented nylon film can be used as a component of high performance laminates, often competing with PET. Nylon is also a good film for metallization since its low thickness allows long runs in the sealed vacuum chamber, giving good production economics.

Nylon has a high melting point and the film is difficult to heat seal, although it will, if corona treated and given enough heat and pressure. Alternatively, it can be coextruded or laminated with easily heat sealable PE.

Nylon is one of the few film materials (PET is another) which in its non-shrinkable form can be used at high temperatures. Therefore, it is used for cook-in bags and vacuum packages for processed meats. For certain applications such as ham and special sausages, the product may be cooked by the manufacturer in the nylon-based package. Nylon film's toughness is useful when vacuum packaging is used; if the product contains sharp particles or bones, its puncture resistance is also a major benefit. For example, nylon is used for packaging hypodermic syringes and military spare parts.

When its gas barrier properties are also required, nylon is a cost effective solution; but if these are not needed, there are less expensive alternatives like HDPE.

Nylon is coextruded and laminated to other film substrates. Applications for coextruded PET/nylon/PE films include bacon, cheese, meat, greasy and oily foods, coffee, gas-flushed products. Nylon is laminated to aluminium foil to make a retort pouch laminate or it can be metallized to improve the barrier and to impart a foil-like look. The metallized material is used in institutional coffee pouches, metallized balloons and (in combination with EVOH and LLDPE) bag-in-box applications.

The high strength and toughness of nylon film can be improved by orientation, which also improves barrier properties and stress crack resistance. Compared with oriented PET film, oriented nylon is a better gas barrier, softer and more puncture resistant, although PET is more rigid and a better moisture barrier.

The most popular oriented nylon film is produced from nylon-6 polymer, and can be either cast extruded as a sheet and then oriented, or extruded in the Japanese 'double bubble' method in which a second, cooler inflation stage takes place on a tubular extrusion line.

The properties of the final material are very much influenced by the degree of crystallinity, which in turn relates to the cooling process. Rapid quenching in the casting process produces an amorphous type of material, whereas slow cooling encourages the more regular crystalline state. Thermoformability and the degree of transparency are also influenced by crystallinity. Since the rate of cooling cannot be as precisely controlled in the tubular blowing process, tubular extrusion usually produces film with lower transparency and gloss.

In thermoforming, nylon can be highly elongated in deep draw moulds and resists stress cracking during moulding. Nylon-6 is frequently coextruded with polyolefins or used as a coating for paperboard, paper and foil, to enhance their properties.

Nylon is one of the most expensive packaging films. The largest sources of oriented nylon films are Italy, the USA, Denmark and Japan.

Other special forms of nylon are finding packaging applications by virtue of their gas barrier properties. Amorphous polyamide (AMPA) can be dispersed within a larger proportion of polyolefin. Immiscible globules of nylon are formed into fine platelets that are dispersed, somewhat like tiles on a roof, to provide a series of barriers to gases, forcing them to take at best a very convoluted path. As a result of continuing pressure for the use of single-polymer materials (to facilitate recycling) interest in this type of mixed material has diminished in recent years.

Amorphous polyamides are also miscible with copolyamides, polyester and EVA. They improve oxygen and aroma barrier and mechanical properties in coextrusions used for bottles, tubes, thermoformed structures and sausage casings.[8]

Although most polyamides have a marked sensitivity to water, AMPA is said actually to improve its oxygen barrier performance at higher relative humidities. In this respect it is exactly the opposite to EVOH, one of its rivals in the high barrier market.

Another polyamide development is MXD-6, a high barrier nylon material based on metaxylenediamine xylene copolymer. This is used in Japan as the barrier core layer of a three-layer PET bottle for wine and beer. Some new polyamides (semi-crystalline as well as amorphous) have also been offered as a sandwiched barrier layer in refillable polycarbonate bottles.

11
Cellophane (regenerated cellulose film)

It is over 100 years since cellophane (originally a DuPont tradename for regenerated cellulose film, or RCF) was patented in the UK. The name combines the words cellulose and *diaphane*, the French word for transparent. It was a very expensive transparent film, available in beautiful sheer colours, used for the packaging of luxury items. In the 1930s the technology was developed for coating it with a moisture-protective layer.

For 40–50 years the material enjoyed a steady growth in packaging applications. At the time, it was the only such clear packaging material. However, once oriented polypropylene film became available, its properties so closely matched cellophane that it rapidly began to be adopted as a lower-cost substitute. So closely are the two materials identified today that they are frequently referred to jointly as comprising the Cellopp market.

Cellophane is in essence now a niche material catering to speciality markets where its characteristics are unique. It maintains a 'dead fold' (keeps its shape after folding) which is important for twist-wrap applications like hard candy wrapping. It is easy to tear, which makes packs easy to open. It is easy to cut and seal, has a high level of gloss, resists high temperatures, and its high moisture permeability can be an advantage for products like cheese and pastry which require protection against bacterial growth. When coated, cellophane is a good barrier to moisture and oxygen. Coatings also enable the material to be heat sealed.

Worldwide consumption for 1995 was estimated to be 1.45 million tons with the few large producers located in the USA,

Mexico, Europe, Russia, China and Japan.[9] The primary markets for cellophane are for confectionery twistwrap, foods like pastries and soft cheeses, pharmaceuticals and healthcare products. The overall ratio of cellophane to OPP in Europe is generally thought to be about 1:6 on a weight basis and about 1:10 if calculated by area since cellophane at 1.5 g/cm^3 is considerably denser than OPP at 0.905 g/cm^3.

Cellophane is considered a polymeric material since it comprises long chain molecules of repeat units, but it is not a thermoplastic because it is neither produced by a melt phase nor capable of being shaped by heat.

The material is produced from high purity wood pulp (eucalyptus is especially suitable) by dissolving the cellulose fibres in carbon disulphide, then adding sodium hydroxide which converts the solution into viscose, a dissolved wood pulp. The gelatinous material is 'ripened' for a few days and then removed through a narrow slit orifice onto a casting drum on which it then passes through further stabilizing liquids, mainly sulphuric acid. This regenerates the film by coagulating the viscose solution. After passing through various washing baths, the material is plasticized, to make it less brittle and more usable as a packaging material, by adding ethylene glycol or propylene glycol (only the latter material is allowed for this purpose in the USA).

At this raw stage the material is extremely moisture sensitive and for the vast majority of applications it is then coated, using either nitrocellulose or PVdC barrier/heat-seal lacquers, the latter providing much better performance. As a result, there are four primary grades of cellophane (with minor deviations in different countries):

P = plain, uncoated film

MS = moisture-proof nitrocellulose-coated film

MXDT = coated with PVdC on one side

MXXT = coated with PVdC on two sides

The MXXT material is the most important and the highest performing grade. It is further subdivided by reference to the method used for coating the PVdC layer. MXXT/A has an aqueous dispersion coating and MXXT/S has a solvent dispersion coating. The A grade has a slightly better barrier. Unlike most other flexible packaging films, cellophane is specified not by thickness but by gauge (g/10 m^2) ranging from 280 to 600 g/10 m^2.

Although cellophane has lost many of its traditional overwrapping applications to OPP, it continues to be used as a laminate in combination with LDPE, BOPP PVdC and/or metallized polyester. Such laminates are used primarily to package pretzels, popcorn, chips, nuts, meats and cheeses.

Cellophane is a versatile material: it can be dyed in a range of attractive colours and is an excellent substrate for the metallization process. Combining these two techniques produces some dazzling visual effects.

The environmental effects of cellophane are a matter of debate. It is produced from renewable natural resources, and the uncoated form is biodegradable. However, high quality wood is used, which is less 'renewable' than other woods, and most applications are coated, and so they biodegrade very slowly. It has been demonstrated that cellophane can be made from other cellulosic feed materials including straw, but this is more difficult and expensive. If the material could be produced economically from recovered wastepaper sources, and the nature of the volatile and liquid waste products associated with its manufacture could be made more acceptable, it could lay claim to being one of the most 'green' of packaging materials.

Cellulose acetate

Cellulose acetate is one of the derivatives of cellulose which has been called a very early form of plastic. It has some excellent properties including very high transparency and gloss. It is easily converted into cartons, fitments and window patches by thermo-forming, adhesives and solvent welding.

Further derivatives, cellulose acetate propionate and cellulose acetate butyrate, have generally similar properties but greatly improved toughness.

Cellulose acetates have been replaced in most major applications by other plastic materials. Transparent cartons are produced more and more from high clarity calendered PVC, especially with the advent of better creasing techniques which overcome PVC's tendency to whiten and fracture at the creases, as well as from APET sheet and polypropylene. The use of cellulose acetate as thin film for window patches in cartons and envelopes has come under threat from oriented polystyrene film which is easier to manufacture and is in more plentiful supply.

12
Plastics' barrier properties and performance

The preceding chapters have described the most common packaging materials and their properties. This chapter shows how the materials compare in general, one with another, in strength and barrier properties. It also describes some new materials that are finding packaging applications based on their high performance.

Barrier property comparison

Two of the most important properties for food packaging are low water vapour transmission rate and low gas permeability. This is because dry food products require protection from moisture and most foods require protection from oxidation.

It is easy to find a plastic to provide a good water vapour barrier, but the lack of a good thermoplastic oxygen barrier is flexible packaging's most critical performance limit.

Some plastics, like polyolefins, are good water vapour barriers but leak oxygen like a sieve. Others, like nylon MXD-6 and EVOH, are good oxygen barriers but are moisture sensitive.

PVdC excels at both oxygen and moisture resistance. PET and PVC rank in the mid-range on both properties.

At the risk of oversimplification, Tables 12.1 and 12.2 show the barrier properties of the commodity packaging plastics. They are listed in order of increasing permeability (the best barriers are listed first), illustrating the lack of correlation between water and oxygen barrier.

It should be noted that permeability also varies with temperature and humidity conditions.

Table 12.1 Water vapour transmission rates of selected polymers[a]

Polymer	WVTR (nmol/m s)
Vinylidene chloride copolymers	0.005–0.05
High density polyethylene (HDPE)	0.095
Polypropylene	0.16
Low density polyethylene (LDPE)	0.35
Ethylene vinyl alcohol, 44 mol % ethylene[b]	0.35
Polyethylene terephthalate (PET)	0.45
Polyvinyl chloride (PVC)	0.55
Ethylene vinyl alcohol, 32 mol % ethylene[b]	0.95
Nylon-6.6, nylon-11	0.95
Nitrile barrier resins	1.5
Polystyrene	1.8
Nylon-6	2.7
Polycarbonate	2.8
Nylon-12	15.9

[a] At 38°C and 90% RH unless otherwise noted
[b] Measured at 40°C
Source: Delassus, P 'Barrier polymers' *The Wiley Encyclopædia of Packaging Technology* (1997) p 74

Table 12.2 Oxygen permeability of selected polymers

Polymer	Oxygen (nmol/m S GPa)
Vinylidene chloride copolymers	0.02–0.30
Ethylene vinyl alcohol copolymers	
dry	0.014–0.095
at 100% RH	2.2–1.1
Nylon-MXD-6[a]	0.30
Nitrile barrier polymers	1.8–2.0
Nylon-6	4–6
Amorphous nylon (Selar[b] PA 3426)	5–6
Polyethylene terephthalate	6–8
Polyvinyl chloride	10–40
High density polyethylene	200–400
Polypropylene	300–500
Low density polyethylene	500–700
Polystyrene	500–800

[a] Trademark of Mitsubishi Gas Chemical Co.
[b] Trademark of E I DuPont de Nemours and Co., Inc.
Source: Delassus, P 'Barrier polymers' *The Wiley Encyclopædia of Packaging Technology* (1997) p 74

As mentioned in Chapters 5–11, there are variations of each plastic, like orientation or metallocene catalyst technologies, that improve the basic polymer's barrier and strength properties. It has also been shown that combination materials, like coextrusions, laminations and blends can optimize the properties of each material. Chapter 13 explores these issues as well as coatings and other surface modifications such as metallization and silicon oxide deposition which can improve a plastic's barrier properties.

Other high barrier plastics — HNPs and fluoropolymers

Two other high barrier materials have been introduced to packaging. High nitrile polymers are superior oxygen barriers and fluoropolymers are superior water vapour barriers.

High nitrile polymers (HNPs) are copolymers of nitrile and other plastics. Nitrile alone is an outstanding gas and aroma barrier and has good chemical resistance, surpassed only by PVdC and EVOH. Nitriles were used in the first plastic bottles for carbonated beverages because of their barrier properties.

HNPs, however, have an affinity for water, and are not a good water barrier. Furthermore, nitrile alone is difficult to melt-process because it tends to degrade at temperatures below those required for processing, and so it is copolymerized with comonomers which increase its melt-processability without reducing its properties.[10]

HNPs can be copolymerized with many different polymers (ABS and SAN, described earlier, are styrene/HNP copolymers), but most of the packaging applications have involved polyolefins. They gain gas barrier and chemical resistance from the HNP, and water vapour barrier and economical processing from the polyolefin. When copolymerized with PP, HNP can be used in a high temperature environment like a microwave. HNPs are stiffer than PET, PVC and the polyolefins.

Rubber-modified acrylonitrile-methyl acrylate (AN/MA) copolymer (Sohio Chemical tradename Barex) is the first HNP in commercial production which has been approved in the USA for use with foods. There have been concerns with other HNPs regarding the migration of the AN into food products.

Barex is used to make chemical-resistant bottles; HNP is the inner contact layer, most often coextruded with HDPE. It is used in blow moulding and stretch blow moulding processes. Injection blow moulding is used mainly for producing small bottles for products like typists' correction fluid and solvent. Larger bottles for chemicals are extrusion blow moulded. Stretch blow moulding improves impact strength.

Barex-based film is coextruded or laminated to polyolefins and aluminium foil for food packaging applications. In semi-rigid sheet form, such coextrusions are thermoformed to make meat and cheese packaging. Since it can be sterilized by either ethylene oxide or gamma irradiation, it is increasingly used for medical product packaging.

Fluoropolymers are a class of paraffinic polymers that have some or all of the hydrogen replaced by fluorine. Although there are several fluoropolymers available, there is only one used for packaging, a modified polychlorotrifluoroethylene (PCTFE) fluorpolymer, tradename Aclar or Kel-F.

Aclar is the best water vapour barrier polymer available. It is transparent, a good barrier to gases (surpassed only by EVOH), is inert to most chemicals, resists abrasion and weathering, and retains its properties over a wide temperature range, from cryogenic to 150°C.

Aclar film can be heat sealed, printed, thermoformed, metallized and sterilized. It is generally laminated to another material. The greatest use is in lamination with PVC for pharmaceutical blister packs, where a high barrier to moisture is required to maintain efficacy. It is also used for packaging moisture-sensitive military, electronic and aerospace items where its high price can be justified.

Mechanical strength comparison

Polymer strength increases with increasing molecular mass and with increasing intermolecular forces, but it decreases in the presence of plasticizers. This is why, at the same molecular mass, nylon and polyester are stronger than polyolefins, and plasticized PVC is weaker than rigid PVC.

Table 12.3 compares the strength of the commodity packaging plastics, on the basis of tensile strength and percentage elongation at break, impact strength and tear strength. They are listed roughly in order of decreasing strength (the strongest are listed first).

Tensile strength indicates the stress that a material can resist before breaking when stretched. For the same amount of material, PET has the greatest tensile strength by far of all the commodity plastics, and LDPE has the lowest. Polycarbonate (described in the next section), nitrile and oriented polypropylene also have high tensile strength.

The lower density polyethylenes and nylon are the most stretchy materials, indicated by the measure of percentage elongation at break. This is one reason why they are used for stretch film. HNP, HDPE, OPP and PVdC are the least stretchy, most brittle plastics.

Impact strength is the material's resistance to breakage under a high-velocity impact. Polycarbonate, PET and PVC have the highest impact strength, which is one reason why they are used for soft drinks bottles. The polyolefins have lower impact strength.

Tear strength combines tensile, shear and elastic properties to indicate the force necessary to propagate a tear. The polyethylenes have the highest tear strength. OPP, which 'zippers' when torn, has a very low tear strength.

Table 12.3 Typical plastic mechanical properties

Type of plastic	Tensile strength psi	Elongation %	Impact strength kg/cm	Tear strength g/0.001 in.
PET	25 000–33 000	120–140	25–30	13–80
Polycarbonate	10 000	92–115	100	16–25
Nitrile	9 500	5	High	High
OPP	9 000–25 000	60–100	5–15	4–6
PVdC	8 000	40–100	10–15	10–20
Nylon	7 000	250–500	4–6	20–50
Fluorocarbon	5 000–10 000	50–400	2–15	3–4
Ionomer	3 000–5 000	350–450	6–11	15–150
HDPE	3 000–7 500	10–500	1–3	15–300
LLDPE	3 500–4 500	500–700	8–13	80–800
MDPE	2 000–5 000	225–500	4–6	50–300
PVC	2 000–16 000	5–500	12–20	Varies
LDPE	1 000–3 500	225–600	4–6	50–300

Source: Soroka, W *Fundamentals of Packaging Technology* IOPP, USA (1995) pp 11–22

Other high performance plastics

There is a family of high performance engineering polymers, some of which have minor uses in packaging. The most notable is very strong polycarbonate. This section briefly describes it and some new high-temperature materials, polyurethanes and liquid crystal polymers.

Polycarbonate (PC)

Polycarbonate, as seen in the strength comparisons in the preceding section, is an extremely strong polymer. It is best known for its use as a vandal-proof glazing, police riot shields, crash helmets and sterilizable infant feeding bottles.

PC is clear, tough and heat resistant. It is the most impact resistant of all plastics. It is also expensive. It has good resistance to water, oil and alcohols, but relatively poor resistance to alkalis. The principal manufacturer, General Electric Plastics, has for a number of years promoted its product Lexan for packaging (Makrolon is a competitor produced by Bayer).

Early applications were for returnable milk bottles in the USA during the late 1960s. Up to 100 trips were claimed, and at this level the initial high cost could be justified, but the bottles were not adopted in the USA. Some school systems in the USA have experimented with single-serving (0.25 litre) bottles in their lunch programmes. Trials in other countries, notably in the UK in the 1970s and 1980s, did not lead to wider adoption until the 1990s when renewed interest in the perceived benefits of multi-trip systems for milk led to some adoption in Germany, Austria, Switzerland, Italy and the UK.

PC is used for large refillable water bottles (5 gallon capacity). These bottles take advantage of PC's light weight, impact resistance, excellent optical properties, inertness and the ability to be washed on the existing equipment at 70°C many times. When coextruded with amorphous nylon, returnable PC bottles can be used for carbonated drinks.

PC can be sterilized by commercial sterilization techniques, such as autoclave, ethylene oxide, gamma radiation or electron beam, making it a good material for many medical applications. Since it can withstand high heat, it is also suitable for retort and hot-fill food applications. A bake-in pouch for partially baked bread rolls is one of the best known food packaging uses for polycarbonate.

PC film for packaging is coextruded with a polyolefin heat-seal layer. Such films are tough and strong, and have been used to package disposable medical products. Thicker sheets, coextruded with crystallized polyester, are thermoformed into strong dual-ovenable trays and blister/clamshells for medical products. Bottles, coextruded with EVOH and PET, are used in Denmark for tomato ketchup; they can withstand hot-filling at temperatures which PET alone could not withstand without distortion or whitening. PC can also be foamed to form a strong insulating material.

High temperature plastics

There are some other relatively new plastics that can be used in high temperature applications.

Polyetherimide is a rigid material characterized by its high temperature stability: up to 180°C in continuous use, but some specialized forms can go up to extremely high temperatures, as high as 350°C. General Electric Plastics manufactures it under the name Ultem and has produced, besides electrical components, coextruded sheet materials to be thermoformed for microwave and ovenable trays.

Polyphenylene sulphide and polyphenylene oxide (PPS and PPO) are two other high temperature polymers used mainly for consumer durable items and engineering components. Some marginal applications have been found in packaging too, when their chemical resistance, heat resistance and mechanical strength justify their higher costs. As mentioned earlier, PPO has been combined with PS for microwavable trays.

Methylpentene copolymer (TPX) was developed by ICI in the 1970s as a high temperature coating for paperboard-based ovenable trays. It has temperature stability up to 100°C. The material was not widely adopted owing both to the difficulty in developing suitable coating technologies and to the advent of PET offering the potential to provide a similar high temperature coating at a much lower cost. Mitsubishi has progressed the packaging applications, producing a high temperature film and moulded containers for cosmetics and toiletries. The latter application exploits TPX's particularly high clarity and sparkle. The material's density, at 0.83 g/cm^3, is lower even than polyproplene's, providing a high yield to marginally offset its high cost.

Norylresin, developed by GE Plastics is heat resistant and maintains dimensional stability up to 137°C. It is used for microwavable trays in the USA and Europe.[11]

Polyurethanes

Polyurethanes (PUs) are a group of thermosets that are used mostly in packaging in the form of cellular dunnage void filler and cushioning materials, although there are some PU films. Polyurethanes are created by the chemical reaction resulting from combining an isocyanate with a polyol (polyester, polyether or graft polymers).

There are two types of PU foam used in packaging. The earliest were the pre-formed spongy resilient sheets used to protect small, lightweight items. This is the same kind of low density, open cell, flexible PU foam used in seat cushions and carpet padding.

The second type, foam-in-place systems, combine the isocyanate and polyol at the point of use. The two chemicals are mixed and dispensed into the box, mould or a bag, where they quickly (5–30 seconds) react to create foam and expand to fill the space. As the blowing agent expands in the forming polymer, the cellular structure is created and the foam rises and then sets into the moulded form.

In most applications, the product itself is placed onto the expanding foam (which is usually covered with a polyethylene film) so that the foam can mould itself around the product, fully filling the void. Foam-in-place material is available in various densities, and can be used for various weights of product. The material is expensive, compared with other cushioning and void fillers, but its advantage is that the mould costs are low or non-existent. It is often used for applications where product shapes vary.

Grades of polyurethanes are also used in the form of coating, and for certain special uses, like suspension packaging, the material is produced as a very tough, thin film. A particular characteristic of the film is its soft feel, making it especially suitable for medical and hygiene products. Its mechanical strength and grease resistance have been utilized in certain demanding industrial and military packaging applications.

Graft copolymers, ionomers

Graft copolymers are a group of speciality plastics with useful properties in their semi-molten form. These are derivatives of some of the most common polymers, modified to have grafted onto their main molecule certain other groups (or radicals).

Graft copolymers, by their affinity to bond to otherwise incompatible materials, provide highly aggressive adhesives.

EVA (discussed in Chapter 7) was an early form; it and DuPont's Surlyn can be used as a film or as a coating and/or intermediate layer in a coextrusion. This is a family of materials known as ionomers, characterized by the presence of metallic ions in the molecule. They relate closely to LDPE and the two most important grades are based on zinc and sodium ions in the polymer.

The outstanding property benefits of these over LDPE are greater toughness, oil and grease resistance, hot tack (physical strength during the molten phase) and a tenacious ability to adhere to other surfaces, especially metals. This mix of properties makes them particularly suitable as a surface coating on paper or multi-material laminates. The toughness of the material is best demonstrated by the use of Surlyn film for skin packaging which can cover even sharply pointed items without being punctured.

As a heat-sealable layer, ionomers provide high strength which builds up very rapidly. This high strength is achievable even if the surface is contaminated, a particular benefit for packaging oily food products such as cheese and ham. A common structure for this application is a laminated PET/nylon with an ionomer heat-seal layer.

Other graft copolymers include ethylene butylacrylate (EBA), ethylene methylacrylate (EMA), ethylene acrylic acid (EAA) and ethylene methacrylic acid (EMAA). All can be produced in film form but are most cost effectively used as blends with LDPE, as intermediate layers in coextrusions or as high quality heat-sealable layers.

Liquid crystal polymers (LCPs)

One final group of materials which are still in the realm of research but which could revolutionize plastics is liquid crystal polymers (LCPs). First used for ultra high performance fibres (Kevlar is one), their potential wider application in film and moulded items has been the subject of research since the early 1980s.

There are two main types of LCP. The lyotropic type can be produced only directly from a solution, to be spun into fibres or into film via a slit die. The other group, called thermotropic, holds greater potential interest for packaging. These materials are characterized by a very sharp (i.e., narrow range) melting point. In their highly liquid form, the molecules align themselves to some extent in the direction of flow, so providing directional rigidity. Some people have called them self-reinforcing plastics for this reason. This also contributes to the excellent permeability resistance to gases. By controlling the processing conditions, the directionality of the long chain molecules can be influenced, and it is even possible that by using electromagnetic fields, quite specific molecular alignments may be possible, providing ultimate strength from the materials. Xydar is an example of this type of LCP, which has been used in the USA for a microwavable dish, but one penalty for its high temperature performance is that it needs to be processed at nearly 500°C.

Thermotropic LCP film offers performance advantages compared with high performance polyesters (PET/PEN), having better thermal, mechanical and barrier properties. It can be used for food and medical packaging applications requiring sterilizing, retorting and long shelf-life.

A newer material, for which processing temperatures under 300°C are expected, Vectra from Hoechst–Celanese, is based on a random copolymer of hydroxybenzoic and hydroxynaphthoic acids. High cost still limits the use of such materials but new techniques, once discovered, have a habit of becoming used more widely and finally finding applications in commodity areas like packaging.

LCPs have another interesting property utilized in digital displays for calculators: they become opaque when excited by an electrical impulse. Even this exotic property has been suggested for exploitation in special packaging applications in Japan by C Itoh and Ajinomoto Foods. A clear window of their ACT film, a PET/LCP/PET laminate, can be made opaque when a tiny voltage is applied. The falling cost of disposable batteries and the Japanese fascination with novel forms of packaging explain this conception.

Endnotes for Part 3

1 'Packaging without plastics: more waste and more energy' *Neue Verpackuging* Vol 41 No 2 (1988) p 79

2 Maraschin, NJ 'Polyethylene' *Modern Plastics Encyclopædia 97* McGraw Hill, USA pp B3–B4

3 Simpson, MF and Presta, JL 'Seal through contamination' *Journal of Plastic Film and Sheeting* Vol 13 No 2 (1997) pp 159–77

4 Demetrakakes, P 'New plastic resins search for their niche' *Packaging* (March 1994) pp 25–6; Leaversuch, RD 'Polyolefins tailored for food containers and lids' *Modern Plastics* (May 1997) pp 33–5; Simon, DF 'Single-site catalysts produce tailor-made, consistent resins' *Packaging Technology and Engineering* (April 1994) pp 34–7

5 'Tampoli Co. Ltd establishes metallocene PE film commercial production' *Packaging Trends Japan* No 97 (January 1997) p 6; Yamada, M 'Possibilities of metallocene LLDPE' *Paper, Film and Foil Convertech Pac.* Vol 5 No 2 (1997) pp 35–8

6 Sherman, LM 'Where the growth is in ... rigid food packaging' *Plastic Technology* Vol 43 No 10 (1997) pp 62–3

7 Pidgeon, R 'Plastic beer bottle launch from Bass' *Packaging Week* Vol 13 No 15 (11–18 Dec 1997) p 1

8 'Amorphous polyamide for improved properties' *Neue Verpackuging* Vol 50 No 3 (1997) pp 60, 62, 64

9 Jenkins, B 'Cellophane' *The Wiley Encyclopædia of Packaging Technology* pp 194–5

10 Lund, PR and Sentman RC 'High-nitrile polymers offer thermoforming advantages' *Modern Plastic International* Vol 27 No 2 (1997) pp 75–6, 79

11 'Take away food container is ovenproof' *Food, Cosmetic and Drug Packaging* (May 1996) p 7

Part 4 Composite and ancillary materials

13
Flexibles and other composite materials

Part 3 discussed the principal plastic packaging materials, which can be made either into films or into moulded forms. Each section discussed some applications where at least one plastic material is combined with another material to tailor specific properties.

This chapter covers some of the more common modified plastic materials, including multi-layer structures, surface coatings and treatments, blends, additives and composite structures. There is a special emphasis on flexible packaging applications, especially for food packaging, because most of them use modified materials to achieve a particular set of properties.

Although paper, foil and cellophane are often the substrate for flexible packaging materials, these are generally coated or laminated to plastics in order to improve their heat-sealability, greaseproof or barrier properties. Sometimes the composite material is made from layers of different plastics, each contributing its own unique properties.

Flexible packaging applications are growing, and new materials are stronger and better barriers than ever before. Flexible packaging is particularly well positioned to exploit the increasing opportunities in convenience food markets. Compared with rigid forms, flexibles reduce not only the volume of packaging but also the packaging and transport costs, and result in less mass for disposal.

Overall consumption of flexible packaging materials is forecast to grow at 2.5–3.0% per annum to the year 2001. Table 13.1 shows the trend in western Europe for flexible packaging usage. The apparent relatively slow volume growth forecast is owing to the dominance of PE, for which the annual growth is forecast at barely 1%, while gains made by PP, Nylon and PET are 5% or more. Nevertheless, these tonnage growth figures understate the real gains being made in terms of area of material used for packaging. This is because of the development of ever thinner gauge materials.[1]

Table 13.1 Polymer/film demand for flexible packaging in western Europe, 1995–2001 ('000 tonnes)

	1995	1996	1997[a]	1998[a]	1999[a]	2000[a]	2001[a]
PE	900	909	918	927	936	946	955
BOPP	380	410	437	470	493	524	550
Cast PP	110	114	119	124	129	134	139
Nylon	63	68	72	75	79	84	88
PVC	60	61	62	63	64	65	66
PET	48	50	53	56	58	62	65
Cellulose film	25	25	24	24	23	22	22
Total	**1586**	**1637**	**1685**	**1739**	**1782**	**1837**	**1885**

[a] Forecast
Source: Gaster P *European Market for Flexible Packaging* Pira International (1997) p xiv

Plastic film materials with properties that are enhanced by being combined with another material are also gaining ground for environmental reasons. Besides the fact that modified plastics use material efficiently, after packaging use they are a good source of energy. As incineration with energy recovery becomes more acceptable, there is less call for single material packaging to facilitate conventional recycling.

There is wide scope for modifying the properties of plastics. Four main approaches are possible:

- combining different polymers in multi-layer structures;
- surface coatings and treatments;
- blending or mixing dissimilar materials; and
- incorporating specific additives to promote the desired effects.

Modified plastics generally aim to improve four properties which are important for food packaging: barrier to water vapour and gas (especially oxygen), heat tolerance and cost.

Since most foods need protection from oxidation, the lack of a good transparent oxygen barrier is flexible packaging's most critical performance limit. Aluminium foil, when built into a flexible material, provides an absolute barrier. The main plastic barrier materials (discussed in previous sections) are EVOH, PVdC and amorphous nylon.

The development of an ever-increasing range of sophisticated barrier films in the form of laminations, coextrusions and coatings, including metallizing and silica deposition, has been central to the success of flexible packaging, especially in the food, medical and pharmaceutical sectors.

Progress in this area has been rapid in recent years and further developments are expected to extend the performance and variety of applications, leading to greatly extended shelf-life for a wide range of food products. Research in the area includes the development of films with barrier properties tailored to meet the needs of specific foods, and the development of 'smart' films which can modify their barrier properties in response to changes in temperature and humidity. There is also an effort to develop more films for use in both conventional and microwave ovens.

The second most restricting factor in food packaging is heat tolerance, given the potential importance of such technologies as hot-filling, retort processing and re-heating. Some materials, like retort pouch laminations, can be used for in-pack sterilizing. CPET or polycarbonate can be used if microwave or conventional oven heating is required. Alternatively, the use of newer cool aseptic filling techniques can be used with most plastics. There are also developments with high temperature styrenes and PVC as well as new grades of polyester and copolyester.

Laminates and coextrusions

Multi-layer structures exist because of the desire to combine the best properties of a number of different materials into a structure when there is no single material that will provide the needed performance. For example, PE is economical and a good moisture barrier, but it is a poor oxygen barrier and may stretch too much if used to package a heavy product. PET is a better barrier to oxygen, but is expensive and does not seal well. A lamination or coextrusion of the two materials can result in a strong material with a good barrier at moderate cost.

There are three ways to combine layers:

- adhesive lamination;
- extrusion lamination; and
- coextrusion.

Other thin, non-plastic materials such as cellophane, paper and aluminium foil can also be laminated to plastics. The number of possible combinations is vast.

Most multi-layer structures are based on polyolefins (PE and PP). Polyolefins are low cost, versatile and easy to process. LDPE and LLDPE are valued for their toughness and sealability. HDPE is favoured for moisture barrier and machinability. OPP is chosen for its ability to provide machinable films with high impact strength and stiffness.

The polyolefin base is almost always combined with other resins to improve its properties. Copolymers like EVA are used as outer layers for their low-temperature sealing characteristics. When oxygen, aroma or flavour protection is necessary, aluminium foil, PVdC, nylon and EVOH are used. Printed paper may be included. Other polymers such as PET or polycarbonate may be used as outer layers to provide exceptional thermal or strength integrity.

Multi-layer films are usually made from two or three materials and may include adhesive bonding layers, called tie layers. For example, since nylon and EVOH do not readily adhere to polyolefins, a tie layer is required for multi-layer films that include them.

Most adhesive laminations are made using a dry bond process. A liquid adhesive is applied to one substrate and is dried with hot air. This dried surface is then adhered to a second substrate using heat and pressure.

In the wet bond process, the adhesive is applied to one substrate and then the substrates are joined and dried together in an oven. At least one of the substrates must be porous enough to allow the water or organic solvent to evaporate.

A third adhesive lamination process involves the application of a hot-melt adhesive (mixture of polymers and waxes) to the substrates, joining them as it cools. In the related thermal process, the heat-sealable layer is applied to one substrate and then the layer is heat activated when the two layers are joined.

The adhesives (generally urethanes or acrylics) are chosen to withstand the intended processing and distribution of the product. This may include high temperature retorting, a product with volatile organic migration which may dissolve adhesives, or other packaging components which may react and change colour. Because the adhesives are reactive chemicals that are expected to polymerize and/or crosslink when coated, government food additive regulations control the presence of any unreacted residuals.[2]

Extrusion lamination involves extruding a thin tie layer of plastic (typically LDPE) to bond together layers of film, paper or foil. This is the method used for making aseptic juice package materials, which may have as many as seven layers including paper and foil interleaved and coated with polyethylene.

The advantages over adhesive lamination are lower cost, because no separate adhesive is needed, and there are no environmental emissions. The polyethylene layer provides adhesion as well as adding a barrier in its own right. It permits the processing of thinner layers of film. At present, extrusion lamination accounts for the majority of printed multi-layer flexible packaging.

Coextrusion is the least expensive process. It eliminates entirely the use of separately manufactured substrates and reduces the operation to a single step. Several layers of molten plastic are simultaneously extruded as a single multi-layer material. A schematic of a typical coextrusion process is shown in Figure 13.1. For example, a single-step coextrusion of HDPE and EVA can replace a lamination of OPP and LDPE which would require four steps (PP extrusion and orientation, LDPE extrusion and then lamination).

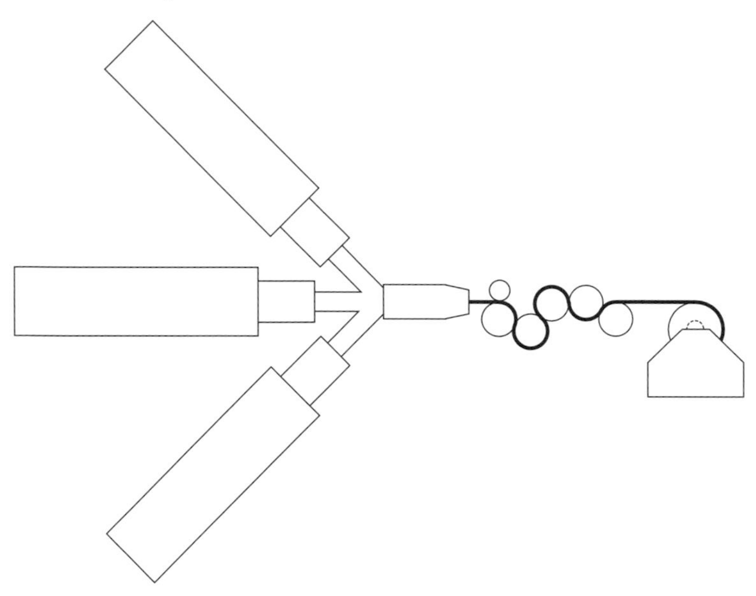

Figure 13.1 Coextrusion

In coextrusions, each plastic maintains its identity as a separate layer and can contribute various functions accordingly. For example, a coextruded film used to package meats may include nylon, EVOH and ionomer, which provide, respectively, heat resistance, oxygen barrier and low-temperature sealing.

For some combinations of materials, there is insufficient adhesion between the two polymers, and a tie layer is used. These tie layers are thermoplastic adhesives which are also coextruded as a distinct layer. Thus many coextruded materials have three layers: one of each of the two desired components and one tie layer.

Structures with four to five layers are common, especially when EVOH is used for a barrier layer. Since EVOH is moisture sensitive, it must be sandwiched between other polymers for moisture protection, and the combinations often require tie layers, as illustrated previously in Table 7.2.

Some combinations that perform very well as coextrusions would not function as blends. For example, a coextrusion of EVA and nylon would be heat sealable and provide an aroma barrier, but if mixed in a blend, the two materials would 'contaminate' each other; the nylon would lose its barrier power and the EVA would degrade at the temperature necessary to melt the nylon.

Bottles and other rigid packages can also be made by coextrusion processes. Coextrusion blow moulding is described in Chapter 5.

There are three types of coextruded structure. Single resin coextrusions have two or more layers of the same resin, but each layer is modified for a special purpose. One layer may be pigmented or recycled resin, sandwiched between virgin materials to control surface quality and machinability, or one layer may have a different coefficient of friction.

Unbalanced coextrusions, often used for form–fill–seal applications, typically combine a functional layer like HDPE with a heat-sealable resin like EVA. For horizontal over-wrapping, a PP surface layer is sometimes desired for its higher thermal resistance. Another application combines cast PP, which has a limited heat-sealing range, with more sealable polyethylene for single slices of cheese. Stretch wrap can be made to be sticky on one side and not the other by coextruding LLDPE with a less tacky material.

Balanced coextrusions have the same heat-sealable resin on both sides of the film. OPP, for example, is increasingly coextruded instead of coated to attain machinable and heat-sealable surfaces on both sides. Frozen food films are typically constructed with an EVA skin for enhanced sealability. Bags are made with a coextrusion of LLDPE for impact strength with LDPE skins to limit elongation.

Surface coatings and treatments

There are three main approaches to surface modification. Most traditional coatings are applied by rollers, extrusion or immersion; these are used to apply coatings of plastic, lacquer or wax. The second approach, vacuum deposition, is used to deposit a very thin layer of metal, metal oxide or silica (glass). The third approach is surface treatment.

Surface coatings and treatments are used to improve water resistance, oxygen barrier properties, chemical resistance or surface appearance. Some cold seal adhesives are also applied by coating, and these are discussed in Chapter 14.

Traditional coatings

The best known traditional coatings are applied by rollers or extrusion. These include thermoplastics, varnishes and lacquers, which are used to give a protected surface. Clay, calcium carbonate and titanium dioxide are used to give smooth, white and highly printable surfaces.

Lacquers and varnishes are applied by rollers to the surface of paper and folding cartons to give a glossy appearance, protect printing and improve abrasion resistance. There are solvent-based and water-based coatings as well as dry varnishes that cure by oxidation. Water-based coatings are gaining popularity because there is less environmental impact (but they require more drying energy, itself an environmental 'cost'). They are usually applied in a printing press.

Extrusion coating is the most economical way to combine thermoplastics with other materials. PE is by far the most commonly used coating material; the range of olefins is so broad that it provides a wide selection of heat-seal and water vapour barrier properties. PVdC and EVOH are the most popular coatings to provide gas and chemical barrier properties as well as sealability. Thermoset coatings are used on pouch structures to protect the surface from heat-sealing damage. Amorphous nylon as an extrusion coating adds thermal stability.

An extrusion coated material is less stiff than a comparable coextruded material. Although, in principle, the use of multiple coatings can provide extremely high performance, the economics and the potential technical difficulties of multiple adhesion make it uncommon.

Wax impregnation is another traditional form of coating, used mostly for paper and corrugated fibre board materials. In the hot dipping technique, wax is absorbed into the surface pores of the substrate. It is becoming less commonly used, however, as polymeric materials are developed which offer comparable or better cost effectiveness.

Newer methods of wax impregnation using rollers allow for differential coating which leaves some areas of the package free from wax in order to facilitate gluing. Since the 1960s, paraffin wax has been mixed with microcrystalline wax, low molecular weight polyethylene or EVA. Some wax-based coatings can be heat sealed but the bond strength is relatively low.

Dip coating can also be used for rigid containers such as PET with PVdC to improve the oxygen barrier. In bottles, this may be carried out either on the small preforms or on the stretch-blown finished container. In the former case the stretch characteristics of the two materials must be very similar, and the adhesion, which is subject to considerable stresses, must be very strong. A similar requirement applies to printing on metal sheets which are subsequently pressed and shaped for three-dimensional containers.

Coatings on glass, mainly for decoration, light barrier or physical protection, are discussed in chapter 3.

Vacuum metallization and silica deposition

In recent years, significant new coating technologies have emerged which are capable of depositing a very thin layer of metal or silicon dioxide (glass) onto the surface of a plastic film or other substrate. Metal and glass deposition brings flexible packaging one step closer to the protection for food afforded by cans and bottles.

Vacuum metallization was patented by Edison in the nineteenth century, and was for decades used on a small scale for the production of hot stamping foils and for the decoration of fitments. The first metallized plastic film was developed in the 1930s and was used for making Christmas tinsel.

The 1970s saw its expansion into packaging using large-scale production plants to metallize thin films. At first the process was used in packaging to provide decorative effects, and coated cellophane was the main material treated. The improvement in oxygen barrier performance was fairly marginal since MXXT is itself excellent in this respect.

Metallized films are still valued for their decorative appeal, but they are much more significant as barrier materials. Metallization gives the base film barrier properties approaching those of foil. Furthermore, metallized films are more flexible than foil laminations and are less susceptible to damage such as cracks and pinholes caused by flexing.

Plastic film is smooth and relatively easy to metallize. Metallized PET, cellophane, oriented polypropylene and nylon are now established as materials in their own right.[3] LDPE, PC and other films, as well as glass, can be metallized, but examples are rare. Paper can also be metallized, which gives it a decorative appearance but does not appreciably improve its barrier properties.

The thin coating of metal is deposited in a high-vacuum process. The substrate is fed through a high-vacuum environment where vaporized aluminium, from a continuously wire-fed crucible, condenses on the surface of the substrate. Figure 13.2 shows a schematic view of a metallizing chamber.

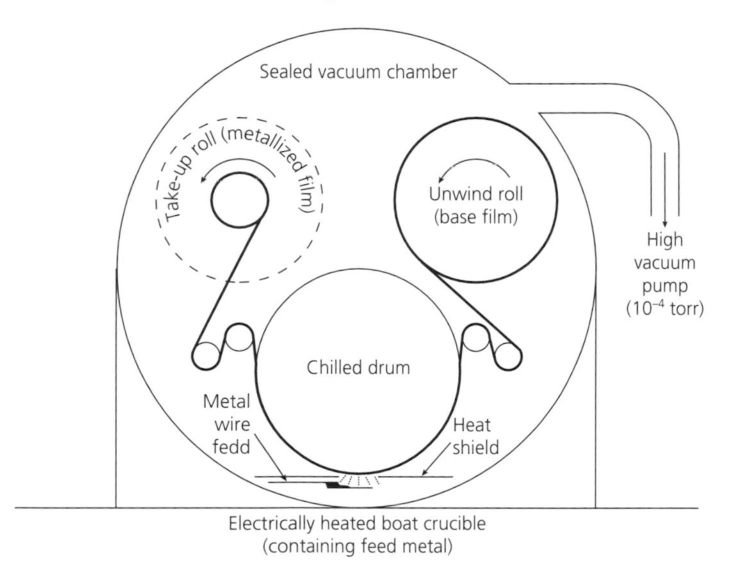

Figure 13.2 Layout of a vacuum metallizer

A significant element of the total cost of metallizing is the proportion of time needed to load the sealed chamber, pump down to the very low vacuum conditions needed, and start the process compared with the actual time taken to run the reel of material past the metallizing head. The length of runs can therefore be optimized by using the thinnest possible films. Oriented PET and nylon, both produced in thicknesses of 12 μm are the most economical in this respect.

Metallized PET films are used for their barrier properties and for abrasion resistance in coffee vacuum packs and snack food packages. Metallized PET has also been used in pouches for electronic components, where it provides static shielding in addition to strength.

The barrier performance of metallized materials varies with the weight of aluminium deposited and its degree of compaction. Factors which affect this are the temperature, distance from the boat crucible to the film, level of vacuum, temperature of the crucible and cooling effectiveness of the chilled drum, as well as the nature of the film surface.

Much effort has been put into establishing some method of defining the amount of metallization, and therefore the barrier quality of metallized materials, using weight of aluminium per unit area, electrical conductivity in ohms per square foot and optical density. The last is the quickest, simplest and is non-destructive, hence it is most commonly used. There is a very clear relationship between the barrier performance and optical density.

Metallized plastics have also been used for the production of microwavable trays and containers, with the intention of promoting high local heat to give products a crisp, dry texture. The phenomenon of metallic particles being excited by microwave irradiation is well known. The metallized susceptor material is usually adhered to a thicker board, like a pizza base, for ease of handling. Temperatures at the surface of the susceptor may reach 200°C. A variant on the use of aluminium for metallizing is the use of vacuum-deposited stainless steel. It is reported that this metal oxidizes at about 200°C, so providing an automatic cut-off to prevent overheating. A whole specialized part of the industry has developed in the USA, providing materials for the microwave sector.

An alternative to the direct metallization process is transfer metallization. This is a large-scale derivative of the hot foil stamping process and was developed to coat very thick materials and paper. These pose difficulties owing to moisture boiling off under the high vacuum conditions, and short run lengths in the case of thick materials. In the transfer process, release-coated carrier webs of thin films (PET or OPP) are metallized and coated with a lacquer. The metallic layer is then transferred to the intended substrate via a heated nip roller and the carrier web can be recoated after cleaning for re-use.

The second area of new surface treatments is in glass (silica) deposition on plastic film to improve barrier properties. It has long been possible in laboratory experiments to coat flat materials with extremely thin layers of inorganic compounds such as metallic and other oxides.

More recently, work has been reported on the direct vapour deposition of silicon oxides (SiO_x) and silicon nitrides by evaporation under high vacuum conditions, in a technique similar to vacuum metallization. The exact formulation of the oxide deposited varies between SiO and SiO_2. Although the latter is the chemically stable form, this is not always achieved, hence the use of the formula SiO_x where x can be measured as something between 1 and 2.

PET is the usual substrate for SiO_x, but PP, PS and nylon have been used. Other materials are more problematic, since SiO_x cannot be evaporated at as high a rate as aluminium without spattering and causing holes in the film. Researchers have discovered that a cold gas plasma-enhanced chemical vapour deposition process shows more promise, especially for heat-sensitive materials like LDPE and PP.

Such glass coatings have the advantage of being excellent barriers, transparent, retortable and recyclable. The widest use of these materials is in Japan for the purpose of packaging liquids like tomato or cream sauce products in stand-up pouches.[4]

An extremely thin layer of silicon oxide, aluminium oxide or aluminium, if it is continuous, can provide an excellent oxygen and water vapour barrier while retaining the flexibility of the thin film (such as PET) substrate. Metallized and silicon oxide coated films, however, have been found to develop pinholes and lose some of their barrier properties when exposed to physical stress like flexing, including during the converting process.

Exposure to high humidity and temperature can also reduce the barrier performance of metallized films if the metal is not adequately protected. Silicon oxide coated materials do not face this danger. Since post-fill retorting of flexible pouches is very big business in Japan, this has provided an obvious advantage for glass coatings. This is an active area of ongoing research.

Plastic surface treatments

A final category of plastic surface modification is based on chemical treatment of the molecules in the accessible surface layer. Some treatments, like chemical priming, etching, cleaning and evaporated acrylate coatings, are used to prepare the surface to be bonded to another material. Others are used to change the properties of the finished plastic.

Best known of these is fluoridation, which converts the surface polyolefin molecules into a range of fluorine-containing compounds of the family exemplified by PTFE (Teflon). This is one of the most inert materials known, and the presence of even a thin surface layer of such materials reduces dramatically the scope for solvent to penetrate into the wall and then permeate through it.

The technique is relatively simple for treating blow-moulded containers: fluorine gas is mixed with nitrogen to provide the blowing air on the extrusion blow moulding machine, treating the whole of the inside surface at the same time. Weight loss by permeation of solvents through the wall of the plastic container is reduced by varying degrees depending on the nature of the solution, but is most improved for carbon tetrachloride, pentane, hexane, heptane and xylene. Packaging for liquid home repair products, including paint, and agrochemicals are major applications.

Flame treatment is used to improve the printability of a plastic film container, or the film's ability to be adhered to another material. The material, or pack, is passed through a flame which polarizes the surface, improving its ink adhesion and ability to be laminated or metallized. A similar approach is corona discharge treatment which is more short-lived.

Plastics additives

Many materials like plasticizers, stabilizers and processing aids are added to plastics in quantities ranging from trace amounts to high percentage levels. The purpose is to improve the performance of the host material by providing an additional property or by suppressing an undesirable tendency.

Mineral fillers such as clay, talc, calcium carbonate and titanium dioxide are also added to plastics to decrease cost. These reduce tensile and impact strength, but can increase stiffness and raise the temperature use range.

This section lists the principal types of performance-improving plastics additives but does not go into the details of their formulations.

Processing aids

Processing aids are used to inhibit heat degradation, improve processing viscosity, and lubricate the plastic flow during processing. There are a number of lubricating agents, which promote the smooth flow of melted plastic over die surfaces and mould cavities. Release agents make it easy to remove plastic items from a mould without changing the plastic's performance.

Heat stabilizers retard degradation from heat processing. Coupling agents improve the adhesion between material layers. Blowing and foam agents are used to produce cellular materials. A recent development makes possible the selective foaming of defined areas within an injection-moulded item after it has left the mould.

Catalysts and nucleating agents control the reactions that give the plastic its properties. These are not, strictly speaking, processing agents, but are integral parts of the plastic manufacturing process. As mentioned in Chapter 6, metallocene catalysts have a growing reputation for being able to produce more uniform PE (and other plastics) with better mechanical and barrier properties.

Antistatic additives are used in PE film to reduce the tendency for web materials to stick to the forming collars of form–fill–seal machinery and also to make it easier to open bags for filling. Antistatic additives are also used to reduce the tendency for plastic materials to attract dust electrostatically.

Mechanical and surface property modifiers

Plasticizers, by a 'lubricating' action at a molecular level, soften rigid polymers and make them more flexible. Impact modifiers are elastomers which improve impact resistance. PVC, which is naturally brittle, usually includes plasticizer and impact modifier.

Antiblocking additives and slip agents prevent film from sticking to itself. They are needed for some grades of LDPE, as are additives to promote the opposite performance, antislip agents, for stretch-wrapping film which is intended to be sticky.

Electrically conductive plastics are used for wrapping some electronics components to avoid damage from electrostatic discharge. Some incorporate carbon to make the film partly conductive, and others use humidity-attracting additives to produce a thin surface film of moisture, to leak away the charge from the surface as it builds up. There is continuing research into ways to make some polymers, like polyacetylene, inherently electrically conductive.

Flame-retardant agents are needed when there is danger of fire. Most plastics are combustible; polyolefins, plasticized PVC and foamed PS are the least flame resistant. However, most flame retardants are toxic, and they should be used only when strictly necessary.

Tiny, thin flakes of mica, a naturally occurring transparent material, can be added to improve barrier performance. DuPont has incorporated the flakes in its Selar barrier resin in an interlay like small barrier tiles; a six-fold increase in barrier performance is claimed to be achieved by a 30% addition of mica to the EVOH barrier. Shell Chemicals has also patented the use of 10–35% mica in a terpolymer for use in multi-layer containers.

Aging modifiers

Since plastics can degrade in the presence of oxygen, ozone, sunlight and biological agents, there are additives to retard the process — or to accelerate it if degradation is desired.

Ultraviolet stabilizers are important because most plastics (like PE, PP, PS, PVC and so on) deteriorate, or photo-oxidize, in sunlight. Ultraviolet absorbers and stabilizers are used to provide protection from this during long-term outdoor exposure. Pigments and titanium dioxide can also provide against ultraviolet rays protection. Antioxidant stabilizers retard atmospheric oxidation and deterioration of the plastic, and are, for example, used widely with PVC to prevent hydrogen chloride elimination. Antizonant agents protect from ozone deterioration.

Biocides, algicide, bactericides and fungicides inhibit biological attack and prolong the life of a plastic.

There are even additives intended to accelerate degradation for litter control or to aid decomposition in a landfill. It should be noted that very little degradation occurs in a modern sanitary landfill. It is also important to note that the two conditions — buried in soil and exposed to sunlight — are mutually exclusive. Photo-initiator additives, like complex metallic additives and carbon monoxide/ethylene copolymers, cause the polymer chains to break when exposed to sunlight. Biodegradable additives, like starch, permit the plastic to be digested by microorganisms when buried in soil.

Optical property modifiers

Colourant agents are used to impart colour to molten plastic. Some are sensitive to temperature, so they must be compatible with the processing temperature. Some include heavy metals (lead, cadmium, mercury, etc.) that are toxic. These are, in the main, the subject of legislation, particularly for packaging intended for food use.

Barriers to ultraviolet light are sometimes required for the packaging of light-sensitive products. In the past, many of these have involved the addition of an amber-coloured compound to filter out the specific harmful wavelength. Recent developments from Japan include Sumitomo's use of fine zinc oxide particles (30–50 nm) and Mitsubishi's use of traces of a naphthalene dicarboxylic acid derivative for the same purpose.

Plastic blends and alloys

The distinction between blends and alloys is not as clear in the plastics world as it is with metals, but this is an area of current development in plastics. The highest growth is predicted in Japan, and the materials expected to feature most commonly are PET, polycarbonate and polyarylate (a heat resistant, UV stable, very tough plastic that is used in automotive, safety, outdoor construction and lighting applications).

As with many of the applications previously discussed, the principal goal is to create high barrier materials, improve heat tolerance or reduce material usage. Multiple grades of a single polymer are often blended; for example LLDPE and LDPE were originally blended for cost optimization when the cost of LLDPE was high.

Some blends involve the use of coupling or making compatible agents which act as dispersed adhesives. Others depend on a natural affinity between materials or act on the simple mechanical trapping of one material in a matrix of another. A particular benefit of blends is that they can be coextruded with polymers that are compatible with only one of the blend's constituents.

Blends containing polycarbonate, polyamide and polyester are the fastest growing.[5] The most noteworthy new blend is PET/PEN, used to make single-serving bottles for soft drinks and beer. The PEN economically improves the temperature performance and barrier properties, over PET, in the small, thin-walled bottle and is expected to take market share from glass bottles and aluminium cans.[6]

Other new blends include: a new oxygen barrier made from a blend of EVOH and LLDPE, which has a much lower cost than the EVOH alone, and is used to package meats[7], and PPO/PS blends for use in retort pouches and ovenable trays.[8] Japanese applications include blends of polyamide added to EVOH and EVOH/PET alloys which provide good gas barriers and overcome some of the problems of using EVOH alone.[9]

Paperboard and metal composite structures

In addition to combination materials based on plastics, there is another group of packages made by combining paperboard and metal with other materials.

The composite can is a good example, combining a spiral-wound paperboard tube (which may have an inner plastic or aluminium layer) with seamed tinplate or aluminium ends. Composite cans can be highly decorated because of the paper, and they can be made impermeable and liquid-tight.

The most numerous of these composites is the range of liquid-tight paperboard cartons which derive directly from the flexible form–fill–seal sector as a material, but owing much to the carton industry in their forms. All of these are based on paper which represents at least 80% of the mass. They are dependent upon at least one thermoplastic layer to provide both sealability and liquid-tightness. Where very high performance is required, these also incorporate a layer of aluminium foil.

A handful of suppliers dominate this market. Tetra Pak from Sweden is the largest, producing the material and aseptic form–fill–seal equipment. The material is either supplied in rolls or pre-cut blanks. The exact material construction used varies between individual companies, partly with respect to the product for which they are used. Although there have been some non-food applications — motor oil is one — the overwhelming majority of these cartons are for liquid foods, mostly milk and aseptic drinks, semi-liquids and pastes.

The most common aseptic construction is PE/unbleached kraft/adhesive/aluminium foil/PE/PE. The two inner coextruded LDPE layers are selected so that the layer in contact with the foil has the highest possible adhesion, and the surface in contact with the food has the least possible taint propensity. Unbleached kraft is used because it is the most cost effective in relation to material strength. Bleaching is costly, reduces strength and is environmentally unfavourable. A white finish is obtained by a clay coating on the surface.

Liquid carton systems can be as large as 2000 ml or as small as 10 ml. They are generally brick-shaped or have gable tops. Some have special dispensing closures like plastic spouts.

A similar carton for dry products is made from paper, aluminium foil and polyethylene. An example is the Cekacan from Akerlund and Rausing, used for sensitive products such as tea, coffee and peanuts.

Another important group of composite containers is thermoformed plastic trays with paperboard panels. The paperboard provides stiffening. Best known of these is the Tritello from Akerlund and Rausing, which is used for margarines and chilled salads. It consists of a pre-cut and scored paperboard blank inserted into a mould where it is fused to a thermoformed shape. It may have an aluminium foil layer to provide a light barrier. In some special applications, aluminium patches can be incorporated to shield from microwave hot spots.

Differing from these mainly in degree are the thin-walled thermoformed pots with an in-mould paper label, where the paper component provides most of the wall's stiffness.

There have been a few three-piece plastic cans developed, made from PET tubes and aluminium ends. However, they have been abandoned because of high cost and perceived non-recyclability. A two-piece PET 'can' with aluminium ends is, however, still in limited use.

Most multi-material packages like these use a minimum amount of material while optimizing the functions of the package. Some, when compared with the bottles or cans that they replace, have a lower impact on air and water pollution, energy consumption and waste volume, and can be disposed of by incineration.

However, multi-material structures have been somewhat discredited in discussions on environmentally friendly packaging because they are not readily recyclable. In Japan, a biodegradable oxygen barrier has been developed to replace an aluminium layer. Silicon dioxide barriers, for example, are very thin and fragment completely when the substrate is recycled. There is urgent demand for this type of packaging in Japan and other countries owing to new packaging recycling laws which make it obligatory to recycle beverage cartons.[10]

14
Ancillary materials

Adhesives

Modern adhesives have come a long way from the animal glue and simple starch-based types first used in packaging. Although starch-based adhesives are still used for a large percentage of applications, they have been gradually replaced during the past 50 years by synthetic materials.

Natural water-based adhesives

Starch-based adhesives, made from starches and dextrins (modified starches), are used for bonding paper. The largest single use is for combining corrugated board. Other uses are for sealing cartons, shipping containers and can labels.

Starch-based adhesives are economical, easy to use and easy to clean up after use. They have excellent adhesion to paper, and they have a good environmental image because they are natural. However, they dry slowly because of their high water content.

Two natural water-borne adhesives are made from animal protein. Animal glue, made from bones and skin, is used to make rigid paperboard set-up boxes and remoistening adhesive for paper tape. Casein, made from milk, is used for labelling glass bottles in high-speed wet applications, for example for beer bottles. It is also easy to remove when bottles are returned.

Natural rubber latex is extracted from rubber trees. It is the only adhesive system that will form bonds only to itself with pressure. It is used in latex cold seal adhesives, which are used to replace heat seals in form–fill–seal applications such as that for confectionery. These materials have made possible major improvements in machine speed, since the small but finite time needed for heat to pass through the thickness of a substrate and fuse the normal sealable layer is no longer required. Furthermore, heat-sensitive materials and products which could not readily be packed on heat-seal equipment are now suitable for wider applications.

Such cold seal adhesives are applied as a coating by the material converter. They provide either non-peeling adhesive joints or, if desired, will peel cleanly by the combined layers of adhesive stripping from one of the substrate surfaces. Particular care is needed in the storage of these materials since they are affected by heat and exposure to air.

Synthetic water-based adhesives

Polyvinyl acetate (PVA) adhesives are the most widely used class of packaging adhesives. One form is the familiar household 'white glue,' safe for use even by children. They are water-based, easy to use and to clean up, and inexpensive.

The properties of PVA adhesives can be tailored for various applications. They are formulated for widely different substrates including paper, glass, metals and some plastics (although water-borne adhesives do not generally work very well on plastics). The largest uses are for case and carton sealing and in spiral wound composite can stock.

PVA adhesives are the fastest-setting class of water-borne adhesives, and so are used in higher-speed production settings. They are tough in heat and cold, and can be made moisture resistant. Their range of usefulness has been extended as newer formulations have become available. Copolymerizing the PVA with ethylene or acrylic esters has greatly improved adhesion, especially to plastics and high-gloss coatings.

Thermoplastic adhesives

Hot-melt adhesives are increasingly being used in high-speed packaging applications where their instant bond and lack of solvents meet the performance required. They are made from a thermoplastic polymer, usually EVA, compounded with a wax or tackifying resin.

Hot melt is the fastest-setting adhesive. Applied in a molten state, it forms a bond as it cools and solidifies. Hot melts can be formulated to adhere to almost any surface, and can even fill gaps. There is no solvent, and they are environmentally benign in the sense of emitting no solvents at the point of use. They are, however, a nuisance in any recycling programme.

Polymers other than EVA can be used. Polyethylene-based hot melts are more economical and are adequate for some paper bonding applications (like case sealing and multi-wall paper bag seaming/sealing). Hot melts based on PP, polyamides, polyesters and block copolymers of styrene-butadiene or styrene-isoprene are used for specialized applications.

Heat-seal coatings, applied to paper, are reactivated by heat (either direct or induced in a metallic layer by radio frequency equipment) in a subsequent operation. They are used to apply inner seals and labels to bottles and jars, and to seal blister materials, envelopes and open-mouth multi-wall bags.

Pressure-sensitive adhesives

The surface of a pressure-sensitive adhesive is a very high viscosity liquid, which provides instant bonding to almost any surface. The primary use is for pressure-sensitive labels and tapes, described in the following sections.

Some solvent-based adhesives, like rubber resins, are still used for pressure-sensitive adhesives, although this is the smallest, and most rapidly declining, class of adhesive. Other adhesives like elastomeric polymers and water-based acrylic have replaced them in most applications for reasons of cost and environmental considerations. For the most demanding applications in flexible packaging the two-part polyurethane adhesive remains the most effective adhesive.

Adhesive tapes

The traditional gummed, paper-based tape with a water-activated adhesive, for so long the mainstay of the case closing market, has gradually been replaced by pressure sensitive plastic tapes for most applications.

Gummed paper tapes can be made very strong in fibreglass sandwich constructions, and can provide excellent fibre bonding strength. However, their effectiveness is very dependent on the care with which they are applied. Surfaces which are too absorptive can dry out the moisture rapidly, leading to a weak bond. Their bond strength also falls off if they are allowed to become too dry before application, or are applied with insufficient pressure.

Pressure-sensitive tapes now dominate the case-sealing market. The most common substrate is biaxially oriented polypropylene, but polyester, unplasticized PVC and paper are also used. Most of the adhesives are based on rubber and a tackifying resin, but there is a trend to greater use of emulsion (water-based) acrylic adhesives.

Personalized printing of pressure sensitive tapes, once possible only for customers willing to order large quantities, is now much more accessible to smaller users. The main reason is that acrylic-based adhesives on PP tapes do not pose the same release problems as rubber-based adhesives on PVC. This, in turn, reduces the need to use toluene-based inks to cut through the release coating, and allows simpler printing inks and equipment to be used, with corresponding savings in cost.

Printed tape can add to the security of sealed cases, and there are a number of refinements to this security element. One is the printing of invisible messages by inks capable of being read only under ultraviolet light. Another, developed in Japan, is based on an abhesive material (to prevent adhesion) printed in selected areas before it is coated with a pigmented adhesive. In use, when the tape is removed, the adhesive separates selectively, leaving a readable message in the pigmented adhesive on the container or case.

Pressure-sensitive filament-reinforced tape has reinforcing fibreglass or other fibrous material embedded in the pressure-sensitive adhesive. It is used where high strength is required, like securing the top layers of a palletload.

Inks

All printing inks consist of a colourant (pigment or dye), a substance that acts as a binder for the colourant, a film former, such as a resin, varnish or polymer, and a liquid dispersant which may be water, solvent, oil or monomer.

Solvent-based inks containing an organic solvent adhere well to all substrates and dry quickly, but like solvent-based adhesives, they produce toxic fumes and are flammable. Printers are no longer permitted to discharge evaporated ink solvents to the atmosphere, and it is relatively difficult to condense them for recovery because of their low evaporation temperature.

Water-based inks have long been used for printing paper and paperboard, and their use is growing because of concern about air pollution. However, water-based inks have drawbacks: they dry more slowly and are difficult to use on plastic surfaces, and even water-based inks contain some solvent or alcohol. These problems are a subject for active research and development.

The viscosity of the ink must be carefully matched to the application. Inks for gravure printing have strong solvents to improve adhesion; such inks would soften the plates used for flexo printing. Gravure inks are thicker than those used for flexo printing. Flexo inks must be able to flow into and out of the tiny engraving on the anilox roll.

It is essential for inks to be at one moment very wet and flowing, and at the next moment dry. The most common drying method for package printing is evaporation of the solvent and/or water. The drying process is costly; it requires time and heat, and causes air pollution. Drying and curing of inks is another active area for research and development.

An important current trend is the rapid acceptance of ultraviolet curing inks. These use ultraviolet light or electron beam (EB) radiation to polymerize and cure, rather than dry, the ink. They are based on polyesters or low molecular weight acrylic formulations that polymerize quickly when irradiated.

The radiation-induced curing is fast, being complete in a fraction of a second while the print is on the press. The inks have better chemical and weather resistance than conventional inks and perform well on a variety of substrates, notably on porous and unevenly coated materials. They are used in flexo, gravure, screen and ink jet printing. UV-cured inks contain minimal volatile materials and do not pose an air pollution risk, but there are safety concerns: workers must be thoroughly shielded from the radiation.

A major requirement for ink jet printing is that the inks must not clog the tiny nozzles in the printer. Ink jet inks resemble writing inks more than they do printing inks; they contain dyes rather than pigments dispersed in water or glycol.

There are a number of new 'reactive inks' with special properties that have found packaging applications.[11] These provide an extra benefit beyond conveying the printed message and graphics. For example, thermochromic inks have been introduced for the purpose of indicating the temperature of a food product. Some have just one colour change point, others have two or more. Some uses are mere novelty like the designs on cups which change when filled with hot or chilled drinks. They are also used in medical packaging to confirm that these have reached sterilizing temperature, and also for time–temperature monitors to indicate food temperature abuse, as described in a subsequent section. Humidity-sensitive inks can be used to monitor climatic conditions for moisture-sensitive products.

An important application for reactive inks is for security purposes. Most are based on exposure to light or a magnetic field. Photochromic inks change colour when exposed to light, and can show if a package has been opened or tampered with. Photo-switchable inks, that change, appear or disappear when flashed by a certain colour of light, offer the opportunity of changing a message to show that some operation, like purchase, subassembly or expiration, has been completed. Invisible (ultraviolet readable) inks can either be broadband readable or can be specially formulated to respond only to precisely defined wavelengths of light.

Magnetic inks are machine readable and could replace optical bar codes. There are also security applications where a magnetic label's field is set by the product manufacturer to trigger an alarm if the product is stolen, and is cancelled out at the store checkout. Magnetic security tags are described in more depth in a subsequent section.

Special security inks can also be formulated using chemical compounds, enzymes and microbiolgical materials such as antibodies which are extremely specific, and can be used to authenticate products which may be copied illegally.

Label materials

The humble label, once little more than a piece of paper stuck onto a pack, is now capable of providing many more functions than merely a carrier of information. The materials from which these are made are also selected from a much wider range of options.

Paper labels

Paper-based labels are still the mainstay of the industry. Paper is usually the most economical label material, and can be printed using any process.

The conventional wet-glue label is widely used for high volume items, but the proportion which are now self-adhesive continues to grow. Self-adhesive labels can utilize permanent adhesives, most often used for package identification, or temporary adhesives that can be easily removed, such as that used for point-of-sale information on kitchenware.

Plastic labels

Although oriented polypropylene is the most common plastic label material, most plastic films and laminates are printable and so may be used as label stock.

The properties of plastic labels are generally modified to give them many of the physical attributes of paper, while avoiding its hygro-instability (frequently a cause of curl on application machines). Cavitated OPP has a paper-like feel, and offers benefits in cost and yield because its density is so low. An additional benefit is the OPP's ability to stretch to a greater degree than paper; this is important where it is used for wraparound labels for PET carbonated drinks bottles where the bottle expands slightly over time and can break a paper label which is unable to stretch. Polystyrene filled with mineral materials such as talc or titanium dioxide also has handling properties very similar to a good quality paper.

Much of the interest in synthetic substrates for labels centres on the subject of plastics recycling. Plastic containers with paper labels cannot be recycled economically since the paper contaminates the recovered plastic. Using plastic labels which are compatible with the container material, for example OPP labels with polyolefin bottles, overcomes this limitation.

Most plastic labels are pressure sensitive. Pressure-sensitive labels are sold affixed to a release-coated (antiadhesive) paper which is peeled off as the label is applied. Though more expensive than regular labels, they eliminate the need for a gluing station, are quick to change over, and are easy to count and track (a special benefit for pharmaceutical packaging).

Most technical developments in recent years have centred on reducing the cost of pressure-sensitive labels. Water-based adhesives, mainly the acrylic type, have been adopted for environmental reasons and lower-cost backing materials have been developed.

Since the release-coated (usually silicone) backing paper is neither recoverable nor recyclable, there have been efforts to develop better release coatings which are applied to one side of backing-less self-adhesive labels. This involves coating the printed surface of the label with a release material such as silicone and leaving the labels in an uncut form on the reel. The reeled labels are similar to printed tape. On a bottling machine, a die cutter is used to punch out the required shapes. This system is most appropriate for applications where a standard shape (not necessarily standard design) of label is used. There have also been attempts to feed pre-cut labels from a block, but this is not widely used.

Special equipment is required to apply shrink labels made from PVC, PET, PS and OPP and stretch labels made from LDPE, PP and PVC. Shrink-sleeve labels were first used in Japan in the 1970s. They take the form of reverse-printed pre-sealed sleeves, slipped over the body of the container and then shrunk (almost exclusively around the girth) by passing through a heat tunnel. Benefits are that very high quality graphics are possible, the print is protected from scuffing, the bottle is afforded some protection from scratching (especially important for glass bottles), and relatively complex shaped bottles can be all-round labelled. By arranging the sleeve to grip over the closure, a tamper-evident feature is also achieved.

Plastic makes it possible for labels to have special decorative textures. PET, PS and OPP are increasingly being used in transparent, frosted, pigmented and pearlescent forms.

Holograms of the type used on labels (or laminated to packaging materials) are produced by embossing an image onto the surface of a substrate using heat and pressure and/or ultraviolet radiation. Some applications use two photographic images which show two moments in an action to for example, illustrate, product use etc.

Unlike traditional printing, no ink is employed in holographic printing. The colours are derived from diffraction of light from the surface; a similar process is employed for embossing diffractive patterns onto packaging materials. Most web-fed packaging materials can serve as the substrate, but OPP is most common.

The other major technical development has been the use of in-mould labelling, in which a printed label coated on the back with a heat-sensitive adhesive is placed inside a mould in which a plastic container is to be produced. The hot plastic causes the label to adhere firmly over its entire area with no loose edges. The system can be used with injection moulding, extrusion blow moulding, injection stretch blow moulding and thermoforming. For thermoforming, a paper label is often used to provide an essential component of the wall stiffness; some yoghurt and dessert packages are produced in this manner.

Substrates for in-mould bottle labelling may be of the usual label materials: paper or foil, but mostly plastics, including the transparent types. It is even more important for these bottle and label materials to be compatible for recycling, since an important part of bottle-making economics is being able to reprocess in-plant scrap. Furthermore, the materials fuse more effectively when they are identical or similar.

Other materials are used for special applications. Soluble papers offer some benefits in returnable bottle systems, while for very high performance situations, such as hazardous chemicals, tough spunbonded polyolefin materials like Tyvek are used. Edible labels made from natural collagen, are used to label meat carcasses. Foil can be printed, but it is almost always necessary to laminate it to paper so that the labels will work in labelling machines.

Smart labels

Smart labels can call home, alert security or read temperature. There are a growing number of labels which are employed for specific tasks in a logistics channel.

Radio frequency identification (RFID) is an automatic data collection method that 'reads' a radio frequency (RF) tag from a distance. RF tags can be read in a variety of circumstances, including not needing a line of sight to the reader and the ability to read multiple tags simultaneously.

The tags are battery-powered and can be passive (read only) or active. The active tags allow the user to program in new information, such as a change in product or location status. They have the ability to carry and update a portable database.

RF transponder tags consist of a microchip and a coil antenna. They can be made with differing signal transmission systems and encoding formats. For example, magnetic-based systems are the most commonly used today, primarily for tagging animals, labelling gas bottles, automobile key identification and factory automation. The range of magnetic systems is limited to a few inches. Electric field and microwave-based systems have a larger range, and are used for rail/toll roads.

Electronic article surveillance (EAS) systems are used primarily in the retail industry to detect theft. A tag attached to the product activates a sensor at the checkout unless deactivated or removed by the retailer.

Three technologies are currently used. Magnetic systems excite a strip of magnetic material which emits an acoustic frequency (acoustomagnetic) or a pulse of energy (electromagnetic) which can be detected. Magnetic and radio frequency systems such as those discussed above are relatively inexpensive and are attached permanently to the goods or their packaging. Microwave tags are more expensive and are removed by the store personnel, but they are reusable.

The tags can be applied by the retailer, the product manufacturer or the package manufacturer. They can be inside the package, on its surface or hanging from it.

Time–temperature indicators (TTIs) are used to monitor the storage conditions of chilled or frozen foods, as well as refrigerated medical products like vaccines and blood. In their simplest form, they are labels printed with low melting point waxes which soften and spread at a fixed temperature. Others cause changes in microencapsulated low melting point waxes. Both kinds provide only a one-off point of reference, showing that the pack has at some time been exposed to a temperature above the threshold.

More sophisticated TTI technologies achieve colour changes using polymerization and enzyme reactions.[12] For example, LifeLines Technology offers self-adhesive labels in the form of a bulls-eye. The sensitive polymer centre, which darkens with cumulative high temperature exposure, is surrounded by an outer reference ring, providing customers with a measure that is correlated with the freshness of a chilled product.

The need for temperature monitoring in distribution is increasing. As food scientists continue to develop atmosphere-modification (described below) and partial preservation techniques, the need for refrigerated distribution and temperature control will continue grow.

Atmosphere-modifying agents

Active packaging material technologies can extend the shelf-life of fresh or partially cooked foods and other sensitive products[13].

They are generally used to control the amount of oxygen or moisture inside sealed packages, or to slow microbial growth. The pack must be totally sealed and have adequate barrier properties, otherwise the agent will simply draw in outside air until it is exhausted.

Plastic films that control the respiration of food are also called active packaging. These films are used in packages to modify (MAP) or control (CAP) the atmosphere inside the package with their selective barrier properties. In CAP/MAP, the headspace is modified by pulling a vacuum, by flushing with the desired gas composition, or by some other atmosphere-modifying technology. For example, many foods are sensitive to oxygen (and yet produce it as they deteriorate), and packing them in oxygen-permeable packaging can reduce the oxygen in the headspace and increase their shelf-life. It should be noted that there is high variability between foods as to the critical level of oxygen, and after a point too little oxygen (conversely too much carbon dioxide) can actually accelerate deterioration and anaerobic pathogen growth. It is important to determine this critical point when tailoring packages to control oxidization.

Desiccants

Desiccants, which absorb water from the air inside a container, are the most widely used atmosphere-modifying agents. Most are made from silica gel crystals packed in small, porous sachets.

A primary use is to protect steel equipment from rusting during export conditions of high humidity for an extended period, inside of a PE-lined crate. Smaller sachets are used in dry pharmaceuticals and foods such as snack crisps and dried milk, where a low moisture content is critical.

A similar function is performed in meat and fish packaging by using a large pad filled with propylene glycol or diatomaceous earth. The pad absorbs water and inhibits spoilage bacteria, increasing the shelf-life by several days.

When used with food products, there should be adequate labelling to indicate that the sachet should not be eaten. In some cases, the sachet is attached to the inner package wall to prevent accidental ingestion. Another example is the use of active alumina incorporated into stoppers for some pharmaceuticals.

Desiccants have a finite absorbing capacity (although silica gel can be redried by heating). Care must be taken when using desiccant (or any other atmosphere-modifying agent) to ensure that the amount of moisture sealed in with the product at the time of packaging does not exceed the absorptive capacity.

Freshness agents

The Japanese have been very active in developing freshness agents. These are chemicals, also packed in small sachets inside sealed packages or attached to the surface of a film, which reactively modify the atmosphere. Three examples are oxygen scavengers, ethylene scavengers and antimicrobial agents.

The most important are oxygen scavengers which absorb oxygen from the headspace to prevent, or at least retard, oxidation reactions which lead to food spoilage. The best known are made from iron filings which rust (inside its sachet) and use up the available oxygen in the process. To eliminate the unsightly rusty appearance which these can sometimes show, and also to prevent rejection problems when metal detectors are used on the packing line, some other chemicals may be employed, such as ascorbic acid (vitamin C). They are used mostly for dry foods since water activity limits their effectiveness.

Similarly, oxidizable polymers have been developed. They have a metallic compound, enzyme, or other scavenger ingredient in the plastic which absorbs oxygen.

Another approach to oxygen control is to use a sachet or film to control (increase or decrease) carbon dioxide. A high carbon dioxide level is desirable for some foods like meat because it inhibits microbial growth. A low carbon dioxide level may also be desirable; for example, a sachet containing a mixture of iron and calcium hydroxide, which scavenges oxygen and carbon dioxide, has been used to package fresh ground coffee in flexible packages, more than tripling shelf-life.

Other freshness agents include ethylene scavengers, which slow the decaying process of fresh produce. For example, sachets containing silica gel with permanganate are used for many fruits, especially kiwis. There are also sachets that transmit microbial inhibitors such as ethanol or sorbate.

Related to these are plastic film materials offered in Japan which contain finely ground silica and other inorganic materials as an odour absorber. Materials claiming this property are sold to consumers for use in refrigerators as well as being used by food manufacturers to eliminate off-odours. At least one Japanese company produces a range of thermoformed trays made from a silica-filled polyolefin for which it claims a ripeness retarding performance.

Microbial spoilage is a surface problem for some foods like bakery, cheese and semi-dried fish products. Antimicrobial agents like ethanol or sulphur dioxide, incorporated in a film or sachet, can be used to deposit vapour onto the food surface, eliminating the growth of moulds and pathogens. For example, a Japanese invention incorporates zeolite, which dissolves silver ions on the surface of the food. The regulatory status of such packaging has yet to be determined.

Volatile corrosion inhibitors

Volatile corrosion inhibitors (sometimes called vapour phase inhibitors) are used to retard rusting and corrosion of ferrous metals, especially for military and engineering items. As with the other atmosphere-modifying agents, it is important to note that the container, such as a crate, needs to have a waterproof seal, and that moisture-containing materials, such as wood, should not be sealed inside the container.

Edible films

Edible films may seem to be an oxymoron, since being part of a food or medicinal product they themselves have to be protected from contamination during handling. Many foods, however, already have a protective coating to prevent moisture loss. Wax coatings on produce, collagen casings for sausages, and shellac or zein coatings on confectionery and pills are some well known examples.

Edible films are expected to have increasing applications for improving the quality of fresh or minimally processed food products. Examples include edible moisture barriers to prevent moisture loss from fresh cut fruits and vegetables, and edible oxygen barriers to stop enzymatic browning. New applications and quantification of barrier properties are areas of active research.

Most applications, like forming a film coating directly on a food or filling a casing with a food or drug, are not intended to eliminate the need for conventional protective packaging (although they may reduce the amount needed). Rather, they are intended to lengthen the product's shelf-life or improve its quality.

Edible films are made from polysaccharides (cellulose derivatives, starch, carrageenan, alginate, pectinate and chitosan), protein (collagen, gelatin, casein, whey protein, corn zein, wheat gluten and soy protein), and lipids derived from plants and animals (carnauba wax, candelilla wax, beeswax, shellac, triglycerides, acetylated monoglycerides, fatty acids, fatty alcohols and sucrose fatty-acid esters).

The films can be cast or extruded. For example, soft gelatin capsules are formed from cast sheets of plasticized gelatin which are sealed and filled in a single operation. Coatings on foods and drugs are generally cast directly onto the product's surface.

Although edible films do not modify the atmosphere of the package, they do control the surface interaction between the product to be ingested and the atmosphere. They, along with the other atmosphere-modifying technologies, can be expected to continually improve the quality and shelf-life of food and drug products.

Endnotes for Part 4

1 Gaster, P *European Market for Flexible Packaging* Pira International (1997) p xiv

2 For example, US Food and Drug Administration *Indirect Food Additives: Adhesive Coatings and Components* Code of Federal Regulations, Title 21, Part 175

3 Kelly, R 'New developments in vacuum coated flexible packaging materials' *IAPRI International Packaging Research Symp.* Reims, France: ESITC (12–14 March 1997)

4 Brody, A 'Glassy coatings' *Asia Packaging Food Industry* Vol 5 No 7 (1993) pp 49–51

5 Stahl, PO and Sederal, WL 'Polymer blends' *Kundstst. Plastics Europe* Vol 86 No 10 (1996) pp 34–6, 1518, 1520, 1522

6 'USA — test marketing PET–PEN bottles' *EU Packaging Report* No 39 (Sept 1996) p 21

7 'Korean film touted for cost, performance' *Modern Plastics International* Vol 26 No 3 (1996) p 24

8 'Packaging seeks out the best materials' *Plastic Rubber Weekly* No 1243 (9 July 1988) pp 15, 24

9 Yoshii, J 'Trends in barrier design' *Packaging Japan* Vol 12 No 63 (1991) pp 30–8

10 'Biodegradable oxygen barrier replaces foil in drinks cartons' *Japan Packaging News* (Jan 1996) p 2

11 Cooper, A 'Developments in specialist processes — security and responsive packaging: intelligent inks' *Package Printing Technologies* Pira Conference Publication (3 Nov 1997).

12 'Is timing right for time–temperature indicators?' *Packaging Strategies* Vol 15 No 22 (1997) pp 4–5

13 Labuza, TP 'An introduction to active packaging for foods' *Food Technology* Vol 50 No 4 (1996) pp 68, 70–1

Part 5 Conclusions and references

15 Material comparisons and conclusions

Each packaging material has its own unique set of properties, and every application has its own set of requirements. It is the skill of the packager to be able to evaluate the market and technical suitability of the wide range of material options available.

This book has described the materials used for packaging and it has highlighted their substitutability. For almost every packaging application, there are competing materials.

The purpose of this final chapter is to provide a general comparison across materials on the basis of market profiles, technical properties and environmental factors.

The comparisons take the form of performance assessment grids (Tables 15.1 to 15.3), listing the most common packaging materials and rating them on various properties. The first version of these evaluations can be found in *Packaging 2005 — A Strategic Forecast for the European Packaging Industry* by R Goddard (Pira International, 1997).

It is important to note that the ratings are based on a simple five-point scale. They are admittedly subjective and the product of the authors' UK/US perspectives. Other writers may well assign different rankings. Furthermore, there are many different forms of each material, some of which may overcome a particular deficiency (metallocene catalyst technology is a good example). The comparisons are intended to be general and qualitative. At the risk of oversimplification, this chapter is intended to consolidate information for broad material comparisons.

It is also important to note that a higher number on the five-point scale always indicates the 'best,' the highest performance. For example, a 5 in barrier performance means a good barrier, and a 5 in cost indicates a low cost, because such characteristics as low weight and low cost are considered to be desirable.

Market assessment of materials

Table 15.1 compares materials based on their market performance. Each variable will be discussed in turn.

Table 15.1 Market profile performance assessment

Material	A	B	C	D	E	F	Total
Paper	2	5	5	5	5	4	26
Paperboard	4	3	5	5	5	4	26
Corrugated	3	4	4	3	5	4	23
Moulded pulp	3	2	4	2	4	5	20
Cellophane	1	1	3	4	3	4	16
LDPE and LLDPE	4	5	5	3	5	3	24
HDPE	4	4	3	3	5	3	22
PP	5	4	4	3	5	3	24
PVC	3	3	3	3	3	1	16
PS	4	2	3	4	4	3	20
EPS	3	1	3	1	4	2	14
PET	4	2	3	4	5	3	21
Laminates	4	2	4	4	3	3	20
Tinplate	3	2	3	4	4	4	20
Aluminium	4	1	3	4	4	4	20
Glass	2	4	2	3	5	5	21

A = Diversity of form B = Cost (1 = highest) C = Distribution efficiency D = Decorative options E = Pan-European suitability F = Environmental perceptions Units = 1–5 by ascending order of importance, by each factor

Diversity of form (A) This represents the versatility of the particular material, including its subforms, in its ability to meet a wide range of market needs. Plastics rate highest, because they can be used to make packages in almost all forms, from films to bottles and other moulded forms. PP rates the highest because it has the most versatility. Aluminium, laminates and cartons also have diverse applications.

Glass, paper and cellophane are limited to traditional forms (glass bottles, paper labels, cellophane film). This, coupled with economics, is one reason why many of their original applications have been switched to plastics.

Cost (B) There are some significant economic differences between materials owing to raw material and fabricating costs, and larger differences owing to the relative density and thickness of container walls made from different materials.

Using these cost criteria, paper and LDPE rate the highest since they are used in many thin, lightweight applications, and their raw material and processing costs are low. Glass and corrugated fibreboard likewise rate high because their raw material costs are low, and HDPE and PP rate high because they are used in thin cross-sections.

The materials which rate poorly on the cost parameter are aluminium (high material cost), EPS (thick cross-sections) and cellophane. The classic substitutability of cellophane verses OPP is a good example of the cost factors: cellophane has a considerable density disadvantage plus a cost premium, and it cannot be produced as thin as OPP. The cumulative effect is that cellophane can never be the economic option where both are equally suitable, which explains why cellophane is now used only in speciality applications.

Distribution efficiency (C) This is related directly to the mass of the package made from each different material, plus the volume and/or performance efficiency with which it can be converted into packs.

Glass, since it is heaviest, rates lowest on distribution efficiency, and freight rate reduction is an important reason why many former glass bottle users have turned to lighter-weight plastics. The materials which rate highest are the paper-based and flexible materials like LDPE and laminates. Flexibles offer two distribution advantages: they are lightweight and thin, and they are shipped to form–fill–seal facilities in rolls, whereas empty rigid containers, like bottles and cans, occupy a larger cubic volume, and represent a greater inbound distribution cost for the filler.

Decorative options (D) These give a measure of the quality of decoration which can normally be applied directly to the material. Separate labels, which may be of extremely high quality, can of course be applied to any form of pack. This measure is related to Table 1.6 in Chapter 1 which shows the possible decorating options for each material.

Paper and paperboard can be the most highly decorated, with effects ranging from colour in the base material to any of the printing forms. Polystyrene and cellophane are easier than most plastics to print and have a high gloss.

Pan-European suitability (E) This is intended to assess how widely each form of material can be used, taking into account local availability of materials and legislative restrictions including recycling mandates. The most common and widely recycled materials, including paper and board, PE, PET and glass, score high on this parameter.

Environmental perceptions (F) Although some of these ratings can be disputed by facts, market perceptions regarding environmentalism may matter as much as, or more than, scientific facts. There is, for example, a perception that glass and moulded pulp are more environmentally favourable than other materials. A later section of this chapter addresses environmental factors from a more scientific point of view.

The exercise of totalling the market profiles provides an artificial result, since the materials are not wholly substitutable, nor are the parameters weighted in any way for which is more important in any given situation. Nevertheless the totals do provide an indication of marketability, and correspond roughly to consumption rates, with paper and paperboard, LDPE and PP in the lead.

Technical performance of materials

Table 15.2 lists a wider spectrum of properties, selected to cover the general usefulness from a technical consideration of each material.

Table 15.2 Technical performance assessment

Material	A	B	C	D	E	F	G	H	Total
Paper	1	2	2	1	4	2	2	3	17
Cartons	2	2	2	2	4	3	3	3	21
Corrugated	1	2	2	2	4	2	4	3	20
Moulded pulp	1	2	2	1	4	3	3	4	20
LDPE and LLDPE	4	4	5	3	2	3	4	3	28
HDPE	4	4	5	3	3	4	4	3	30
PP	4	4	5	3	3	5	4	3	31
PVC	4	4	4	4	3	3	4	3	29
PS	3	4	4	3	2	3	3	3	25
EPS	3	4	4	2	2	2	3	5	25
PET	5	4	4	4	3	4	4	5	33
Other plastics	4	4	4	3	4	3	4	3	29
Tinplate	3	5	4	5	5	3	4	3	32
Aluminium	3	5	4	5	5	4	4	3	33
Glass	5	5	3	5	5	3	2	2	30

A = Chemical resistance B = Water tolerance C = Sealability
D = Barrier performance E = Temperature tolerance F = Forming versatility
G = Strength H = Protection
Units = 1–5 by ascending order of importance, by each factor

Chemical resistance (A) This relates to the level of inertness, or the range of materials which may be safely contained in direct contact. PET and glass rate highest, and they are used extensively for food contact and chemicals. The polyolefins and PVC also rate high.

Water tolerance (B) This includes other aqueous or non-reactive liquids. Glass bottles and aluminium and tinplate cans, used for all manner of liquid foods, tolerate water well; the plastics are not far behind. Paper-based materials do not perform well when wet.

Sealability (C) These ratings reflect whichever technology is appropriate, and judges heat sealing of polyolefins to be best and the sealing of other plastics to be equivalent to the seaming of metal. Paper, which requires adhesive, is the most difficult to seal.

Barrier performance (D) This measure combines gas and water vapour barrier. As explored in Chapter 12, the two properties do not always come together, and some of the best plastic barriers, like EVOH and PVdC, are not considered in this assessment. The best barriers, however, are not plastic; they are glass and metal. The other plastics that perform best on both dimensions are PET and PVC.

Temperature tolerance (E) The two critical points are sterilization temperature, 130°C, and resistance to deep freeze, –20°C. Metals and glass rank highest on the first count, and this is the reason why they are the main materials of choice for retorting and hot-filling.

Forming versatility (F) This reflects the number of different types of material or pack which can be made from a given material. This measure is related to Table 1.5 in Chapter 1 which shows the possible forms for each material.

PP is the most versatile, followed by PET and PVC, because they can all be used as film, thermoforms and bottles, and can all be combined with other materials in a number of ways. Aluminium is versatile because it can be used in cans and trays, and serves as barrier in many laminated structures. The least versatile materials are paper, corrugated fibreboard, glass and EPS, which are limited to their characteristic package forms (paper labels and bags, corrugated fibreboard boxes, rigid containers and EPS cushioning/dunnage).

Strength (G) This relates to how well a pack is able to resist the hazards of distribution and use in the home. No material rated 1 or 5 on this parameter. Each material has its own kind of strength. For example, glass is very strong in static conditions like compression but is easy to break on impact.

The materials that scored highest were the polyolefins, PVC and PET, and the metals. PS scored lower than the other plastics because of its brittleness. Glass and paper scored lowest because they are easy to break or tear.

Protection (H) This relates to the ability of a pack or material to protect its contents from physical hazards. PET, strong enough to survive an impact when full of a carbonated 'explosive' beverage, and EPS foam cushioning material have the best reputation for protection; glass has the worst.

Again, it is not necessarily useful to total the scores across the technical performance parameters, since the importance of the properties varies depending on the specific application. However, it is instructive to note that the materials with the highest scores are the ones that do the hardest work — metal (aluminium rates highest), glass, polyolefins, PET and PVC, followed by PS. Paper-based materials rate lowest, and are used for the least demanding applications, valued more for their graphics than for their technology.

Environmental profiles of materials

Table 15.3 considers several variables related to the environment. They include raw materials inputs as well as disposability.

Materials availability (A) This is the abundance of the raw materials on a worldwide scale. Glass is made from sand, and aluminium from bauxite, which are both abundant, but are at opposite ends of the cost scale because it requires considerably more energy to extract the aluminium.

Paper-based materials are based on wood, a renewable resource. Steel for tinplate and vinyl for PVC are widely

Table 15.3 Environmental performance assessment

Material	A	B	C	D	E	F	G	Total
Paper	4	3	3	4	5	1	3	23
Cartons	4	3	3	4	5	1	3	23
Corrugated	4	2	3	4	5	2	3	23
Moulded pulp	4	3	3	2	5	2	4	23
LDPE and LLDPE	3	3	4	3	5	2	1	21
HDPE	3	3	4	3	5	4	1	23
PP	3	3	4	3	5	4	1	23
PVC	4	3	3	3	3	4	1	21
PS	3	3	3	3	5	3	1	21
EPS	3	3	5	2	5	4	1	23
PET	3	3	3	4	5	4	1	23
Others	3	3	3	2	5	2	1	19
Tinplate	4	4	4	5	1	2	2	22
Aluminium	5	2	5	5	1	3	2	23
Glass	5	2	1	4	1	5	1	19

A = Materials availability B = Manufacturing energy C = Mass
D = Recyclability E = Incineration F = Reusability G = Degradability
Units = 1–5 by ascending order of importance, by each factor

available owing to other uses like automobile parts and building materials. The other plastics are made from the same non-renewable resource, petroleum.

Manufacturing energy (B) This is the level of energy needed to produce the finished material or pack. Aluminium, glass and corrugated fibreboard require the most energy. Plastics and paper are roughly equivalent.

Mass (C) This is the weight of the package. Aluminium, PP and PET rate high on this parameter, in part, because they are used to make the most lightweight packages for carbonated beverages, with glass at the heavyweight end of the scale. Polyolefins are of lighter weight (rate higher) than paper-based materials and other plastics.

Recyclability (D) This relates to how readily the material can be recycled, including any additional environmental costs of doing so. Aluminium and tinplated steel are most easily recycled because they have value (especially Aluminium), are easily separated, and there are already good infrastructures for steel and aluminium recycling.

Paper and glass are also easy to recycle — there is a lot of both, and there are increasingly good infrastructures for them; however, the value of the scrap is lower than for some other materials. PET ranks high because of its high instrinsic value, even though it requires new sorting methods and infrastructure development.

The other plastics, although easy to recycle, have less value and are difficult to separate by type. The worst, from a recycling point of view, are EPS because its low density makes transport expensive, and moulded pulp because it is made from such low grade (albiet already recycled) fibres.

Recyclability also reflects the use to which this package has been put, since any residual food, chemical or other content must be removed thoroughly. This restricts the opportunity to recycle much packaging, for example from pesticides, oil, paint.

Incineration (E) Can it be readily incinerated, and does it allow energy recovery? All of the paper and plastic materials will burn easily and can be used to produce energy. This is why there are many who believe that waste-to-energy incineration is the best 'recycling' method of all: no sorting of plastics or paper types is required and it is a good fuel source. However, glass, aluminium and tinplate cannot be incinerated, and PVC, when burned, can produce toxic gas.

Reusability (F) This represents the material's suitability for refillable or reusable packs. Glass wins this one and holds the record for the longest-running refillable career. HDPE and PP rank high, and they are re-used in many plastic bins and returnable transit packages. PVC and PET also rank high and *could* be re-used, although the former rarely is, but there is considerable activity in developing multi-trip PET containers. Although it is not in the table, PC is also highly reusable in water bottles. Paper is the least reusable.

It should be noted, however, that all packaging materials are re-used more in economies where they have more value. It is not uncommon in developing countries to see small bags made from used newspapers, and useful articles made from used fibreboard, tinplate and woven jute.

Degradability (G) This is the rate at which the material spontaneously degrades in landfill conditions. None of the materials rates particularly high on degradability for a good reason: there is more of a need in packaging for durability when it comes to product protection. Paper products degrade the easiest, and plastics degrade only very slowly in modern sanitary landfills.

Again, adding the totals for environmental factors provides no clear winner, and in this case the totals vary very little. Clearly, some environmental factors are at odds with one another and the importance of the factors varies for different products and distribution systems. The table presents a defensible argument that no one material is clearly superior on the environmental scale to the others.

Conclusion

Packaging is big business. Throughout the world, large quantities of material and other resources are employed in the production and processing of packaging. The use of packaging materials, in turn, facilitates commerce, trade, and makes possible a convenience-oriented standard of living.

The choice of a packaging material is not a simple one. It requires a knowledge of the product, the market and of the range of available packaging materials.

Packaging professionals make hard decisions every day. We have a guilty pride in our technical successes, hoping that nobody holds us responsible for covering the planet with cardboard and polyethylene and broken glass bottles, and wondering if any one will thank us for that extra week of shelf-life.

The best packaging decisions come from a base of knowledge: and our industry has never known more about itself. Information about packaging is available from many sources. This book has provided a basic survey of packaging materials. For more detailed information, the final chapter provides a guide to further sources.

16
Packaging materials library

One of the goals of this book is to provide the reader with sources of detailed reference information on packaging material topics. This bibliography is intended to be a guide to the books which comprise the current literature of packaging, as related to materials.

The references are grouped to correspond to the parts in this book, and many were consulted in producing this book. Additional packaging references can be identified by topic using Pira International's Packaging Abstracts database.

This list does not include the endnotes to each section which document, with current journal article references, specific new developments that may not appear in the more general book literature.

General packaging technology

Aubry, S *Packaging Technology International 1996* Cornhill (1996)

Hanlon, J, Kelsey, RJ and Forcinio, HE *Handbook of Package Engineering* Technomic (3rd edn, 1998)

Institute of Packaging Professionals *Glossary of Packaging Terms* IoPP, USA (1988)

Paine, F *The Package User's Handbook* Blackie (1991)

Robertson, GL *Food Packaging: Principles and practice* Marcel Dekker (1993)

Sacharow, S and Griffin, RC *Principles of Food Packaging* AVI, USA (2nd edn, 1980)

Soroka, W *Fundamentals of Packaging Technology* IoPP, USA (1995)

Part 1: Material selection factors

American Management Association *Packaging and Solid Waste Management Strategies* AMA (1990)

American Society for Testing and Materials *Selected ASTM Standards on Packaging* ASTM (1994)

Ball, R *Integrated Packaging* Pira International (1995)

Brody, AL and Marsh, KS (eds) *The Wiley Encyclopædia of Packaging Technology* Wiley (2nd edn, 1997). Pertinent articles follow:

Bastioli, C 'Biodegradable materials' pp 77–83

Borchardt, JR 'Recycling' pp 799–805

Selke, S 'Environment' pp 343–8

Taggi, AJ and Walker, PA 'Printing: gravure and flexographic' pp 783–7

Doyle, M *Packaging Strategy* Technomic (1996)

Castle, M *Transport of Dangerous Goods: A guide to international regulations* Pira International (1995)

Eldred, NE *Package Printing* Technomic (1993)

Fiedler, RM *Distribution Packaging Technology* IoPP, USA (1995)

Goddard, R *Packaging 2005 — A strategic forecast for the European packaging industry* Pira International (1997)

Harburn, K *Quality Control of Packaging Materials in the Pharmaceutical Industry* Marcel Dekker (1991)

Hine, T *The Total Package* Little, Brown (1995)

Howkins, M *World Packaging Statistics 1997* Pira International (1997)

Institute of Packaging Professionals *Chemical Packaging Committee Chemical Packaging Guidelines* IoPP, USA (1995)

Into the Millennium — Packaging Pira International (1997)

Jenkins, WA and Osborn, KR *Packaging Drugs and Pharmaceuticals* Technomic (1993)

Japan Packaging Consultants *Japan's Packaging Business* Japan Packaging Consultants Corp (4th edn.1995)

Jonson, G *LCA — A tool for measuring environmental performance* Pira International (1996)

LaMoreaux, RD *Barcodes and Other Automatic Identification Systems* Pira International (1995)

Lauzon, C *Decoration of Packaging* Pira International (1992)

Lauzon, C and Wood, G (eds) *Environmentally Responsible Packaging — A guide to development, selection and design* Pira International (1995)

Leonard, EA *Packaging Specifications, Purchasing and Quality Control* Marcel Dekker (4th edn 1996)

Levy, GM (ed) *Packaging in the Environment* Blackie (1992)

Lox, F *Packaging and Ecology* Pira International (1992)

Malthlouthi, M (ed) *Food Packaging and Preservation* Blackie (1994)

McKinlay, AG *Transport Packaging* IoPP, USA (1998)

Mogil, HM (ed) *Packaging Sourcebook*, International Edition, North American Publishing (annual)

O'Brien, JD *Medical Device Packaging Handbook* Marcel Dekker (1990)

Paine, F and Paine, HY *Handbook of Food Packaging* Blackie (2nd edn, 1992)

Perchard, D and Bevington, G *Packaging Waste Management: Learning from the German experience* Perchards (1994)

Pilditch, J *The Silent Salesman*: *How to develop packaging that sells* Business Publications (1961)

Ramsland, T and Selin, J *Handbook on Procurement of Packaging* PRODEC, Finland (1993)

Selke, S *Biodegration and Packaging* Pira International (1996)

Selke, S *Packaging and the Environment* Technomic (2nd edn, 1994)

Stewart, B *Packaging as an Effective Marketing Tool* Pira International (1995)

Stewart, B *Packaging Design Strategy* Pira International (1994)

Twede, D and Parsons, B *Distribution Packaging for Logistical Systems* Pira International (1997)

Part 2: Traditional packaging materials

Bathe, P *Developments in the Packaging of Alcoholic Drinks* Pira International (1997)

Brody, AL and Marsh, KS (eds) *The Wiley Encyclopædia of Packaging Technology* Wiley (2nd edn, 1997). Pertinent articles follow:

Aluminium Association, 'Aluminium foil' pp 458–63

Abendroth, RP and Lisi, JE 'Glass ampules and vials' pp 35–8

Attwood, BW 'Paperboard' pp 717–23

Bayliss, AM 'Multiwall bags' pp 61–6

Caganagh, J 'Glass container manufacturing' pp 475–84

Dixon, D 'Wirebound boxes' pp 113–15

Foster, G 'Corrugated oxes' pp 100–8

Grygny, J 'Molded fiber' pp 382–3

Hambley, DL 'Glass container design' pp 471–5

Irwin, C 'Blow molding' pp 83–93

Johnsen, MA 'Aerosol containers' pp 27–31

Johnsen, MA 'Pressure containers' pp 774–83

Kraus, FJ and Tarulis, GJ 'Steel cans' pp 144–55

Lisiecki, RE 'Gabletop cartons' pp 187–9

Lynch, L and Anderson, J 'Rigid paperboard boxes' pp 108–10

Nairn, JF and Norpell, TM 'Bottle and jar closures' pp 206–19

Norment, RB 'Steel drums and pails' pp 318–24

Obolewicz, P 'Folding cartons' pp 181–6

Quinn, R 'Solid fibre boxes' pp 112–13

Reingardt, T 'Aluminium cans' pp 132–4

Reznick, D 'Can corrosion' pp 139–44

Sikora, M 'Paper' pp 714–17

Silbereis, J 'Metal can fabrication' pp 615–29

Sweeney, FJ 'Barrels' pp 70–1

Waldman, EH 'Molded pulp' pp 791–4

'Wood boxes' (no author given) pp 115–17

Cakebread, D *Paper-based Packaging* Pira International (1993)

Fibre Box Association *Fibre Box Handbook* FBA (1994)

Grikitis, K *Developments in the Packaging of Soft Drinks* Pira International (1992)

Hogan, PM 'Evaluating new coatings for glass containers' *Glass Industry* Vol 67 No 12 (1986) pp 14–16

Institute of Packaging Professionals *Corrugated Container Design, Testing and Specification* (seminar videotape) IoPP, USA (1992)

Jonson, G *Corrugated Board Packaging* Pira International (1993)

Maltenfort, G *Corrugated Shipping Containers: An engineering approach* Jelmar, USA (1990)

Maltenfort, G (ed) *Performance and Evaluation of Shipping Containers* Jelmar, USA (1989)

Paper in Contact with Foodstuffs Pira International Conference Proceeding (1997)

Plaskett, CA *Principles of Box and Crate Construction* US Department of Agriculture (1930)

Roth, L *The Packaging Designer's Book of Patterns* Van Nostrand Reinhold, USA (1991)

Part 3: Plastic packaging materials

Ashby, R, *et al. Food Packaging Migration and Legislation* Pira International (1997)

Berins, ML(ed.) *Plastics Engineering Handbook* International Thompson (1991)

Bigg, DM 'The newest developments in polymeric packaging materials' *IoPP Technical Journal* Vol 10 No 3 (1992) pp 24–36

Blackstone, B (ed.) *Plastic Package Integrity Testing: Assuring seal quality* IoPP (1995)

Briston, J *Advances in Plastics Packaging Technology* Pira International (1993)

Brody, AL (ed.) *Modified Atmosphere Food Packaging* IoPP, USA (1994)

Brody, AL and Marsh, KS (eds.) *The Wiley Encyclopædia of Packaging Technology* Wiley (2nd edn, 1997).
Pertinent articles follow:

Brasington, RM 'Rigid plastic boxes' pp 110–12

Brighton, T 'Stretch film' pp 434–45

Carter, R 'Injection molding' pp 503–11

Carter, S 'Polyethylene, high density' pp 745–8

Coco, DA 'Poly(vinyl chloride)' pp 771–5

DeLassus, PT *et al.* 'Vinylidene chloride copolymers' pp 958–61

Doar, LH 'Heavy duty plastic bags' pp 60–1

Dodge, P 'Rotational molding' pp 819–22

Ferguson, N 'Corrugated plastic' pp 285–7

Finson, E and Kaplan, SL 'Surface treatment' pp 867–74

Firdaus, V and Tong PP 'Polyethylene, linear and very low density' pp 748–52

Foster, RH 'Ethylene-vinyl alcohol copolymers (EVOH)' pp 355–60

Giacin, JR and Hernandez, RJ 'Permeability of aromas and solvents in polymeric packaging materials' pp 724–33

Gibbons, JA 'Extrusion' pp 370–8

Harstock, DL 'Styrene-butadiene copolymers' pp 863–4

Hernandez, R 'Polymer properties' pp 758–65

Huss, GJ 'Microwavable packaging and dual-ovenable materials' pp 643–6

Irwin, C 'Blow molding' pp 90–3

Jolley, CR and Wofford, GD 'Shrink film' pp 431–4

Keith, H 'Foam trays' pp 933–7

Kirk, AG 'Plastic film' pp 423–7

Kong, D and Mount, EM 'Nonoriented polypropylene film' pp 407–8

Kopsilk, DR and Fisher, E 'Plastic bags' pp 66–70

Luise, RR 'Thermotropic liquid-crystalline polymers' pp 69–72

Lund, PR and McCaul, JP 'Nitrile polymers' pp 669–72

Mabee, MS 'Aseptic packaging' pp 41–5

Maraschin, NJ 'Polyethylene, low density' pp 752–8

McKinney, L, *et al.* 'Thermoforming' pp 914–21

Mihalich, J and Baccaro, LE 'Polycarbonate' pp 740–2

Miller, RC 'Polypropylene' pp 765–8

Mont MM and Wagner, JR 'Oriented polypropylene film' pp 15–22

Neumann, EH and Sison, E 'Thermoplastic polyesters' pp 742–5

Newton, J 'Oriented polyester film' pp 409–15

Nurse, RH and Siebenaller, JS 'High density polyethylene film' pp 405–7

Robertson AB and Habermann, KR 'Fluoropolymer film' pp 403–5

Rosato, DV 'Thermosetting polymers' pp 924–7

Singh, RP 'Time–temperature indicators' pp 926–7

Suh, KW and Tusim, MH 'Foam plastics' pp 451–8

Tubridy, MF and Sibilia, JP 'Nylon' pp 681–5

Van Beek, HJG and Ryder, RG 'Film, rigid PVC' pp 427–31

Wagner, PA 'Extruded polystyrene foam' pp 449–50

Wagner, PA and Sugden, J 'Polystyrene' pp 768–71

Wininger, J 'PETG sheet' pp 827–30

Young, WE 'Heat sealing' pp 821–7

Butler, TI and Veazy, EW (eds) *Film Extrusion Manual: Process, materials and properties* TAPPI (1992)

Coles, R *Flexible Retail Packs* Pira International (1996)

Coles, R *Rigid Plastic Containers (Retail)* Pira International (1992)

Cramm, RH and Sibbach, WR (eds) *Coextrusion Coating and Film Fabrication* TAPPI (1983)

Davies, J *Food Contact Safety of Packaging Materials 1990–1995* Pira International (1996)

Demetrakakes, P 'New plastic resins search for their niche' *Packaging* (March 1994) pp 25–6

Edwards, D *Packaging of Pesticides and Potentially Hazardous Chemicals for Consumer Use* Pira International (1995)

Farber, JM and Dodds, K (eds.) *Principles of Modified-Atmosphere and Sous Vide Product Packaging* Technomic (1995)

Flexible Packaging Association *Flexible Packaging Technical Test Procedures and Specifications* FPA (1991)

Florian, M *Practical Thermoforming: Principles and applications* Marcel Dekker (1987)

Grunewald, G *Thermoforming: A plastic processing guide* Technomic (1997)

Hernandez, RJ 'Food packaging materials, barrier properties and selection' chapter 8 in *Handbook of Food Engineering Practice* Valentus, K; Rotstein, E and Singh, RP (eds) CRC (1997) pp 291–360

Hernandez, RJ and Gavara, R *Methods to Evaluate Food/Packaging Interactions* Pira International (1997)

Hotchkiss, JH and Risch, SJ *Food and Packaging Interactions — Vol II* American Chemical Society (1991)

Jenkins, WA and Harrington, JP (eds) *Packaging Foods with Plastics* Technomic (1993)

Jenkins, WA and Osborn, KR *Plastic Films: Technology and packaging applications* Technomic (1992)

Katan, LL (ed.) *Migration from Food Contact Materials* Blackie (1996)

Koros, WJ (ed.) *Barrier Polymers and Structures* American Chemical Society (1990)

Mathlouthi, M (ed.) *Food Packaging and Preservation* Blackie (1994)

Modern Plastics Encyclopædia 97 McGraw-Hill, USA (1996) and *Modern Plastics Encyclopædia Handbook* McGraw-Hill, USA (1994)

'Polyolefins get tough with metallocenes' *Plastics Technology* Vol 42 No 8 (1996) pp 40–2

Selke, SEM *Understanding Plastics Packaging Technology* Hanser, Germany (1997)

Sherman, LM *Plastics in Contact with Foodstuffs* Pira International (1996)

Simmons, B *Recycling of Plastics Packaging — An update* Pira International(1994)

Simon, DF 'Single-site catalysts produce tailor-made, consistent resins' *Packaging Technology and Engineering* (April 1994) pp 34–7

Part 4: Composites and ancillary materials

Brody, AL (ed.) *Modified Atmosphere Food Packaging* IoPP, USA (1994)

Brody, AL and Marsh, KS (eds) *The Wiley Encyclopædia of Packaging Technology* Wiley (2nd edn, 1997).
Pertinent articles follow:

Alsdorf, MG 'Extrusion coating' pp 378–81

Bakish, R 'Vacuum metallizing' pp 629–38

Bassemir, RW and Bean, AJ 'Inks' pp 511–14

Butler, LL 'Tags' pp 875–9

DeLassus, PT, *et al.* 'Vinylidene chloride copolymers' pp 958–61

Dembrowski, RJ 'Coextrusions for semirigid packaging' pp 240–2

Dunn, TJ 'Multilayer flexible packaging' pp 659–65

Eubanks, MB 'Composite cans' pp 134–7

Fairley, MC 'Labels and labelling machinery' pp 536–41

Finson, E and Kaplan, SL 'Surface treatment' pp 867–74

Hatfield, E and Horvath, L 'Coextrusions for flexible packaging' pp 237–40

Hill RJ 'Film, transparent glass on plastic food packaging materials' pp 445–8

Idol, RC 'Oxygen scavengers' pp 687–92

Kannry, H and Latto, G 'Waxes' pp 962–4

Kaye, I 'Adhesives' pp 23–5

Krochta, J 'Edible films' pp 397–401

Mallik, D 'Holographic packaging' p 492

McKellar, RW 'Gummed tape' p 883

Nissel, FR 'Flat coextrusion machinery' pp 231–4

Nugent, FJ 'Vacuum bag coffee packaging' pp 948–9

Perdue, R 'Vacuum packaging' pp 949–55

Rooney, ML 'Active packaging' pp 2–8

Rosato, DV 'Plastics additives' pp 8–13

Schiek, RC 'Colourants' pp 242–56

Sheehan, RL 'Pressure sensitive tape' pp 883–7

Tanny, SR 'Extrudable adhesives' pp 25–7

Wright, WD 'Tubular coextrusion machinery' pp 234–7

Chamberlain, M *Marking, Coding and Labelling* Pira International(1997)

Eldred, NR *Packaging Printing* Pira International (1993)

Farber, JM *Principles of Modified-atmosphere and Sous Vide Product Packaging* Technomic (1995)

Finlayson, KM *Plastic Film Technology: High barrier plastic films for packaging* Technomic (1989)

Gaster, P *European Market for Flexible Packaging* Pira International (1997)

Hall, I *Labels and Labelling* Pira International (1994)

Institute of Packaging Adhesion Technical Committee *Adhesives in Packaging: Principles, properties and glossary* IoPP, USA (1995)

Institute of Packaging Electronic Survelliance Packaging Committee *Electronic Surveillance Packaging: An outline of the state of the industry* IoPP, USA (1997)

Labuza, TP 'An introduction to active packaging for foods' *Food Technology* Vol 50 No 4 (1996) pp 68, 70–1

Miller, A *Converting for Flexible Packaging* Technomic (1994)

Parry, RT (ed.) *Principles and Applications of Modified Atmosphere Packaging of Foods* International Thompson (1993)

Rooney, ML (ed.) *Active Food Packaging* International Thompson (1995)

Tag and Label Manufacturers Institute *Glossary of Terms for Pressure Sensitive Labels* TLMI (1992)

Willhoft, EMA (ed.) Aseptic Processing and Packaging of Particulate Foods Blackie (1993)

Index

Figures and tables in *italics*